SpringerBriefs in Environment, Security, Development and Peace

Volume 2

Series Editor

Hans Günter Brauch

For further volumes:
http://www.springer.com/series/10357

Mely Caballero-Anthony
Youngho Chang · Nur Azha Putra
Editors

Rethinking Energy Security in Asia: A Non-Traditional View of Human Security

 Springer

Editors
Mely Caballero-Anthony
Centre for Non-Traditional Security (NTS)
 Studies
S. Rajaratnam School of International
 Studies
Nanyang Technological University
Singapore 639798
Singapore

Nur Azha Putra
Energy Studies Institute
Energy Security Division
National University of Singapore
Block A, #10-01
29 Heng Mui Keng Terrace
Singapore 119620
Singapore

Youngho Chang
Division of Economics
Nanyang Technological University
14 Nanyang Drive
Singapore 637332
Singapore

ISSN 2193-3162   ISSN 2193-3170   (electronic)
ISBN 978-3-642-29702-1   ISBN 978-3-642-29703-8   (eBook)
DOI 10.1007/978-3-642-29703-8
Springer Heidelberg New York Dordrecht London

Library of Congress Control Number: 2012938952

Printed on acid-free paper

Springer is part of Springer Science+Business Media (www.springer.com)

Preface

In 2008, the Centre for Non-Traditional Security (NTS) Studies at the S. Rajaratnam School of International Studies, Nanyang Technological University, Singapore, convened a conference on *Energy and Non-Traditional Security*. This volume represents selected papers from the conference, updated by the authors to reflect recent developments.

The volume was first conceptualised against the backdrop of the world oil crisis of 2007–2009 and the impacts of the steep price rises on human collectivities. The period of the global oil crisis also saw the world experiencing a number of other security concerns: food security, infectious diseases, water scarcity, to name just a few. At the same time, the global community also began to grapple with the complex threats brought on by climate change—rising temperatures, increasing frequency of torrential rains, as well as flooding and devastating natural disasters that inflicted large-scale damage to property and infrastructure, and displaced thousands of people.

These threats compel us to rethink security: what is security and for whom? These questions are not new. It has been recognised that security as a concept within the field of international relations and international security has evolved over time especially following the end of the Cold War. Today, security can no longer be primarily viewed through the traditional lens, that is, military threats to the sovereignty of the state alone. Instead, the international community has recognised that threats to the state are mostly non-military in nature.

The global food price crisis and the international financial crisis from 2008 to 2010, and global outbreaks such as severe acute respiratory syndrome (SARS) in 2003 and H1N1 pandemic influenza in 2009, are recent examples of how non-military threats could inflict more harm to states and people than traditional threats such as inter-state wars and conflicts. The recent case of the Fukushima nuclear disaster in Japan in 2011, which was triggered by a stronger-than-usual earthquake and tsunami, illustrates how even the most prepared societies can be overwhelmed by the onslaught of complex disasters.

Yet, these types of threats—now referred to as NTS threats—had often been regarded as less important than the threats of inter-state wars, conflicts and violence. These threats had commanded less attention, and had thus had less resources allocated to dealing with them, than traditional military threats. Notwithstanding the disparity in importance that these issues may have received in the past, NTS concerns such as energy security and the complexity of challenges that they present are now starting to change security debates and are galvanising efforts to come to a greater understanding of how they threaten the welfare and security of peoples and states.

Recent work in this field defines NTS threats as "challenges to the survival and well-being of people and states that arise primarily out of non-military sources ... these dangers are often transnational in scope, transmitted rapidly due to globalisation and technological advances" (Caballero-Anthony 2007, p. 1). A key element in the study of NTS is the need to re-examine traditional approaches to dealing with these issues and to analyse how new, innovative approaches can be adopted in dealing with transnational challenges. The NTS lens also examines the role of the state and other actors and stakeholders in dealing with NTS challenges. The NTS perspective in dealing with emerging security challenges therefore requires one to revisit assumptions and to question the premise on which an issue becomes a security threat. This exercise is most useful in the case of understanding energy security and teasing out the security dimensions that come along with it.

This volume begins with the chapter by Mely Caballero-Anthony, Swee Lean Collin Koh and Sofiah Jamil in "Rethinking Energy Security: A Non-Traditional View of Human Security". The chapter synthesises and highlights the importance of bringing in non-traditional perspectives on energy security. Today, energy security is no longer just about the security of supply. Environmental concerns and socioeconomic development have now become inextricably linked to questions of energy security. The authors argue that the new energy security landscape demands a comprehensive approach that includes multi-actor engagement.

Lina A. Alexandra in "The Role of Indonesia's Civil Society in Energy Security" explores how non-state actors such as civil society could be a vital component in dealing with energy crises. She observes that there is awareness among government and civil society actors in Indonesia of the importance of working together in the planning and implementation of the country's national energy policy. However, collaborations do not always materialise as envisioned, with the result that civil society organisations are left at the periphery in terms of their role in any government initiatives. Alexandra concludes by suggesting several strategies that could help civil society organisations improve their political clout and efficiency.

Youngho Chang and Nur Azha Putra in "The Non-Traditional Security (NTS) Perspective on Energy Security Policies in Singapore" examine the way in which Singapore's policies have been shaped more by the imperatives of economic development than human development. They argue that the country's energy policies could be enhanced to improve human security by transforming consumers into active participants in the country's newly liberalised national energy market. They suggest that provisions, in the form of technological capabilities and changes to legislation, should be made by the authorities to allow households to trade their surplus electricity in the open market, which would have the effect of empowering these households and increasing energy efficiency.

In the chapter on "Perspectives on India's Energy Security", Krishnan Rekha unpacks India's energy security policies by analysing the country's strategic response to the energy challenges arising from its rapidly expanding economy and population. The situation is exacerbated by India's growing reliance on fossil fuel imports and a decline in its national natural gas production. Rekha observes that India's response has been focused on transforming its domestic energy consumption and production to achieve greater efficiency; streamlining its energy-related bureaucracies with the intent of fostering greater communication, cooperation and understanding with one another; and also improving India's participation in the global energy system.

Yuxin Zheng and Sofiah Jamil in "Beyond Efficiency: China's Energy-Saving and Emission Reduction Initiatives vis-à-vis Human Development" examine China's efforts in addressing its energy challenges, and its achievements. The authors suggest that more has to be done to realise energy savings and emissions reductions. A deepening and broadening of the concept of energy intensity would be needed. The government would also have to review and amend existing regulations on energy use, and extend energy-saving measures. Further, the country has to explore development paths other than industrialisation.

In sum, this volume on *Rethinking Energy Security: A Non-Traditional View of Human Security*, together with its companion volume on *Energy and Non-Traditional Security (NTS) in Asia*, has aimed to widen the debate on energy security beyond the conventional views of what energy security means to the security and well-being of states and societies. We hope that with these two volumes, a more robust debate can be encouraged to allow for a more comprehensive approach in dealing with energy security challenges. We also hope that in advancing an NTS perspective to this complex security challenge, we can meaningfully contribute to the promotion of human security in Asia and beyond.

Mely Caballero-Anthony

Bibliography
Caballero-Anthony M (2007) Non-Traditional Security and Multilateralism in Asia: reshaping the contours of regional security architecture?, Policy Analysis Brief. The Stanley Foundation, Muscatine, June 2007

Abbreviations

NTS   Non-traditional security
SARS   Severe acute respiratory syndrome

Contents

Chapter 1
Rethinking Energy Security: A Non-Traditional View of Human Security

Mely Caballero-Anthony, Swee Lean Collin Koh
and Sofiah Jamil

Abstract This chapter highlights non-traditional security (NTS) challenges in relation to energy security. It first frames energy security within the context of human security, which provides the foundation for NTS, and then highlights the NTS issues that have arisen in recent years, such as the traditional energy sector (i.e. fossil fuels) and alternative sources of energy. Finally, it attempts to provide some recommendations for addressing these issues keeping the future role of markets, governance, civil societies and technology in focus.

Keywords East Asia · Energy · Environment · Human security · Non-traditional security · Nuclear

M. Caballero-Anthony (✉) · S. Jamil
Centre for Non-Traditional Security (NTS) Studies,
S. Rajaratnam School of International Studies (RSIS),
Nanyang Technological University (NTU),
Block S4, Level B4, Nanyang Avenue,
Singapore 639798, Singapore
e-mail: ismcanthony@ntu.edu.sg
URL: http://www.rsis.edu.sg/about_rsis/staff_profiles/Mely_Anthony.html;
www.rsis.edu.sg/nts

S. Jamil
e-mail: issofiah@ntu.edu.sg
URL: www.rsis.edu.sg/nts

S. L. C. Koh
S. Rajaratnam School of International Studies (RSIS),
Nanyang Technological University (NTU),
Block S4, Level B4, Nanyang Avenue,
Singapore 639798, Singapore
e-mail: iscollinkoh@ntu.edu.sg
URL: www.rsis.edu.sg

M. Caballero-Anthony et al. (eds.), *Rethinking Energy Security in Asia:
A Non-Traditional View of Human Security,* SpringerBriefs in Environment, Security,
Development and Peace 2, DOI: 10.1007/978-3-642-29703-8_1, © The Author(s) 2012

1.1 Introduction

Much has happened in the last five years that has had significant impacts on energy security—from a financial recession and oil/food crisis in 2008 to the trials and tribulations of moving forward in addressing global climate change and, most recently, the implications of the earthquake and tsunami that hit Fukushima, Japan, in March 2011. The various political, socioeconomic and even environmental dynamics at play in the above-mentioned events have demonstrated that energy security is not merely a matter of securing a steady supply of energy sources but also one that necessitates the existence of a complementary environment that is conducive to sustainable human development. This is particularly evident from projections of increased global population and energy consumption in the coming decades, especially in industrialising countries such as China and India. Developing Asian countries have had to reconsider their options for secure energy sources in order to sustain and support their economic growth and development. What is clear from the multitude of developments is that balancing the demands of various stakeholders has become increasingly more complex, thereby making it difficult for countries to break free from their strong dependence on limited energy sources. As such, the energy security concept needs to also include environmental and socioeconomic impacts, the resultant interdependent nature of which necessitates a rethink of the role of markets and governance—one that transcends the national level to involve non-government actors and greater international cooperation.

Focusing on the Asia-Pacific region, this chapter aims to illuminate contemporary energy security issues with particular emphasis on the non-traditional security (NTS) challenges of energy security, For instance, its socioeconomic and environmental impacts.[1] It first frames energy security within the context of human security, which provides the foundation for NTS, and then highlights the NTS issues that have arisen in recent years—first, the traditional energy sector (i.e. fossil fuels) and second, alternative sources of energy. Finally, it attempts to provide some recommendations by looking at the future role of markets, governance, civil societies and technology for addressing these issues.

1.2 Energy Security as a Non-Traditional Security Issue

Unlike traditional definitions of security, which revolve around the stability and security of the state in terms of territories and military capabilities, 'human security' highlights security from the perspective of individuals and communities

---

[1] This chapter brings into discussion many of the points and reflections raised at the Regional Workshop on Energy and Non-Traditional Security, Singapore, 28–29 August 2008. The workshop was conducted by the Centre for Non-Traditional Security Studies, S. Rajaratnam School of International Studies, Nanyang Technological University, Singapore.

within the state. Human security plays an important part in human development, as highlighted by the late Dr Mahbub ul Haq in the Human Development Report 1994, which was published by the United Nations Development Programme (UNDP) (UNDP 1994). In the report, he defined human security as a broadening of the scope of global security to include seven threats: economic security, food security, health security, environmental security, personal security, community security and political security.

At first glance, energy security, as set forth in these categories, would not seem to have any part to play in human security. Energy security as a concept was born out of the 1973 oil crisis and was primarily focused on finding ways to handle any disruption of oil supplies from producing countries (Yergin 2006). This traditional definition remained the primary emphasis for energy security until newly emergent energy-related environmental and socioeconomic issues rose to prominence in recent years, which brought to the forefront elements related to a non-traditional paradigm of energy security in categories of human security such as economic, environmental and food security.

In 2003, the UN Commission on Human Security (CHS) further enhanced the concept of human security by highlighting its two pillars: (a) the freedom from want; and (b) the freedom from fear. From these two pillars, it is evident that energy security would come under the first pillar, given the constant need for energy despite its increasingly limited availability and the challenges being faced in ensuring the safety and sustainability of existing energy sources.

1.3 Securing Traditional Energy Sources via a Non-Traditional Security Lens

When considering NTS-related concerns, a discussion on contemporary energy issues would inadvertently demonstrate sources and implications of energy that may be broader than traditional. This section will examine several NTS concerns that have arisen from the energy sector in recent years.

1.3.1 Growing Energy Consumption Amid Rising Carbon Emissions

A primary dilemma facing many countries, as highlighted in other chapters, has been ensuring a steady flow of energy sources for sustaining development while addressing growing concerns of mitigating climate change where traditional energy sources, such as fossil fuels, play significant roles with their carbon content. Environmental changes, such as the melting of ice glaciers and consequently rising sea levels and coastal flooding, have been clearly documented in the Fourth Assessment Report of the Intergovernmental Panel on Climate Change (IPCC).

Fig. 1.1 World energy demand, 2005–2030. *Source* EIA (2008, p. 1), at: http://www.eia.gov/forecasts/archive/ieo08/pdf/0484(2008).pdf (22 September 2008). Abbreviations: *Btu* British thermal unit, *OECD* organisation for economic co-operation and development

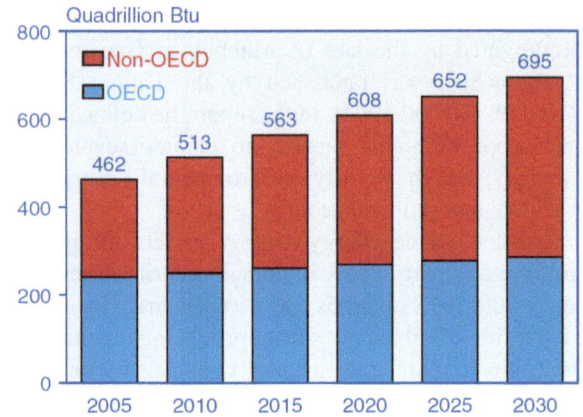

In South and Southeast Asia, for instance, an estimated 1 million people have been placed at risk from flooding along the coastal regions, with negative impact on the infrastructure as well as the aquaculture (Ardiansyah 2008).

Yet, despite such global environmental concerns, demand for fossil fuels is projected to continue to grow. According to the International Energy Agency (IEA), global energy usage increased by 23 % between 1990 and 2005, with a corresponding 25 % increase in greenhouse gas (GHG) emissions (IEA 2008, p. 15). For the same period, global electricity demand increased by 54 %, and oil products were the majority element in the total energy mix at 37 % (IEA 2008, p. 15). IEA also reports that the global oil demand was expected to reach 87.7 million barrels per day (mmbd) in 2009, constituting an annual increment of 1 % from the 86.9 mmbd seen in 2008 (IEA 2008, p. 4). The urgency is further heightened by a reference scenario proposed by the IEA according to which the world's primary energy demand is expected to grow by 55 % between 2005 and 2030 (IEA 2007; see Fig. 1.1). Another significant projection is the expected change in coal consumption, which is estimated to increase from 25 % in 2005 to 28 % in 2030 (IEA 2007). In part driven by rising energy demands in rapidly growing economies, such as China and India, the increase in coal consumption would also drive up GHG emissions over the same time period, with the bulk of emissions being generated by developing countries instead of industrialised ones (see Fig. 1.2). Energy security in the future is therefore likely to be confronted by not only the projected increase in demand and consumption of primary energy, such as oil and coal, but also the environmental consequences of a parallel increase in GHG emissions.

1.3.2 Insecurity from Energy Price Volatility

The stability of energy markets is also an important factor in ensuring energy security. Energy has been central to international economic development since the industrialised era, in particular for the developing nations. This is especially true for East Asia, where the onus of nation building has been placed on socioeconomic

Fig. 1.2 Projected energy-related CO$_2$ emissions by region till 2030. *Source* IEA (2002, p. 73), at: http://www.iea.org/weo/docs/weo2002_part1.pdf (22 September 2008). Abbreviations: *CO$_2$* carbon dioxide, *OECD* organisation for economic co-operation and development

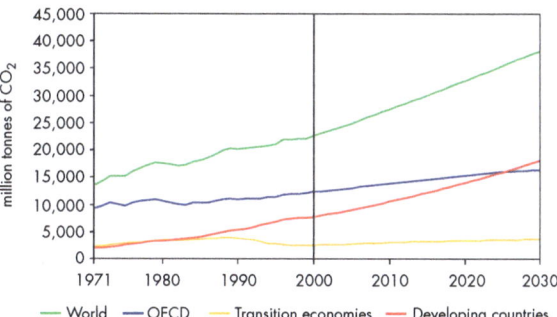

development since the last world war. Governments in some of these countries are prone to be adversely affected sociopolitically by fluctuating energy prices, as was evident from the serious implications of price fluctuations in 2008—high oil prices in July 2008 not only made it difficult for economies to function and remain productive but also brought about other highly acute problems for communities, one of which was higher food prices as a result of higher production costs.

After the peak in oil prices, late 2008 marked the beginning of the global financial crisis, which dealt a further blow to economies worldwide as oil prices began to decline. The Organization of the Petroleum Exporting Countries (OPEC) even set out to reduce oil production by as much as 1.8 mmbd in response to the declining energy prices,[2] sparking future supply security concerns. With the fall in oil prices, several countries in the region contemplated the reinstatement of fuel subsidies. However, such a measure would have encouraged excessive energy consumption and wastage among consumers, who being unexposed to its true scarcity value would have little incentive to conserve energy. Fuel subsidies would also reduce the incentives for industries to introduce energy efficiency measures or adopt alternative fuels (Mendoza 2008).

This is also the case for the Indonesian government, which has the lowest local fuel prices in the region and is thus vulnerable to energy price fluctuations and concerns related to political expediency. An earlier announcement to reduce subsidies by the government, for instance, resulted in widespread street protests.[3] In response to the current energy price dip, the Indonesian authorities are once again contemplating a further reduction of subsidised fuel prices.[4] As before, this

---

[2] With oil prices continuing to plummet, it might be inevitable that OPEC would contemplate further production cuts in order to drive up energy prices; see "OPEC Decides Cut in Oil Production, US for "Free Market" Policy—Iran Paper", in: *BBC Monitoring International Reports* (31 October 2008).

[3] Further fuel subsidy reductions would not come, at least until the end of elections in 2009, although protests have ceased in Indonesia; see "Fuel-price Protests a Test for Jakarta Govt", in: *The Straits Times* (14 May 2008).

[4] Pump prices could potentially decrease by 15 %, with an anticipated encouragement for increased consumption, which could beset efforts to develop cleaner alternative fuels; see "Update 1—Indonesia Energy Ministry to Propose Fuel Price Cut", in: *Reuters* (4 November 2008).

is likely to encourage increased consumption and impede efforts to explore cleaner alternative fuel technologies. Still, fuel subsidies remain a politically sensitive issue; governments might risk straining their financial resources in the long term, as energy prices continue to fluctuate, while debilitating their ability to sustain or extend the subsidies and diverting scarce resources away from more tangible socioeconomic development programmes. The impact of such initiatives could prove even more adverse for financially less-endowed nations, especially under the veil of the global financial crisis.

The fall in energy prices has stifled research and development (R&D) of cleaner alternative fuels, thus causing an even bigger setback for renewable energy sources.[5] The global financial crisis is also projected to take its toll on present and future energy production programmes, which require an annual investment of US$360 billion to meet demand.[6] All these mean that, despite energy prices having fallen from the July 2008 peak of US$145, there is simply no space for relief for governments, energy firms and consumers, who bear the brunt of any increases in prices.

1.3.3 Insecurities from Traditional Energy Exploration Projects

In a bid to overcome vulnerability and dependence on oil imports, countries such as China and Indonesia have sought to utilise their own sources of fossil fuels. For China, a stronger push in the traditional energy sector was also its way of overcoming the economic downturn. The stimulus plan that China put in place to recover from the 2008 financial crisis led to greater electricity use. In the first three months of 2010 for instance, power consumption rose 24.2 % compared to the same period in the previous year, and the engines of China's economy, such as the industries of steel, non-ferrous metals and petrochemicals, recorded new highs of electricity use.[7]

This, to some extent, has left NTS concerns, which can be grouped into two broad areas, on the backburner (Jamil et al. 2010). First is the security of workers, as many workers in these projects are lowly paid, have little cover under protection schemes and are highly susceptible to health and safety hazards. Mismanagement and the lack of oversight only serve to exacerbate the potential for mishaps occurring in energy resource exploration and extraction projects. The disastrous oil spill off the Gulf of Mexico in 2010 is a case in point. Moreover, in the absence of

---

[5] The drop in energy prices essentially makes continued reliance on traditional energy more appealing while clean energy development becomes an uneconomical endeavour given the reduced leverage due to falling oil prices; see "Energy: The Dawn of a Disturbing New Reality", in: *Financial Times* (3 November 2008).

[6] "Energy: The Dawn of a Disturbing New Reality", in: *Financial Times* (3 November 2008).

[7] Peng, Nie, "Power Use by Energy Guzzlers Hit a New High", in: *China Daily* (22 April 2010), at: http://www.chinadaily.com.cn/business/2010-04/22/content_9762990.htm (22 December 2011).

proper and adequate central government purview, the implementation of local energy projects in China in order to feed its rising energy demands has been fraught with problems of socioeconomic marginalisation, such as discrimination against migrants displaced by dam construction projects, reported enslavement of workers and deaths in unsafe coal mines (Andrews-Speed and Ma 2008). A notable example is that of Yumen City in China's Gansu Province, where the local government was heavily reliant on oil revenues from the Yumen oilfield. It became increasingly indebted as production declined, which eventually resulted in its inability to provide vital public goods and a deterioration of economic conditions in the province (Andrews-Speed and Ma 2008). Second is the sustainability of existing energy exploration projects, as oil spills and coal tailing spills have highly detrimental effects on surrounding areas. Aside from contributing to health and environmental degradation, the lack of proper management of such projects can also result in the area being uninhabitable for communities living nearby.

Such concerns gain increasing relevance in light of the efforts of countries to explore not only other potential fossil fuel sites but also alternative energy development projects, such as nuclear energy, and this will be discussed in the following section.

1.4 Energy Security in the Asia-Pacific: What are the Alternatives?

The Asia-Pacific is now regarded as an oasis of socioeconomic rise, spearheaded by the giant economies of China and India.[8] Rising affluence as a result of socioeconomic progress in the region is expected to lead to a corresponding increase in power consumption. As such, to date, Asia-Pacific governments continue to stress the importance of energy supply security for sustaining economic growth. There is nothing anomalous here given the current level of underdevelopment of alternative energy sources that can viably substitute fossil fuels on a mass scale. As such, governments in the Asia-Pacific remain hugely concerned about being able to secure fossil fuels, at least until a time when alternative energy sources are able to account for a significant portion of their national energy portfolios. This persistent and established concern over energy supply security—especially with respect to fossil fuels—is reflected in the quest of countries for emergency oil measures. China, Japan and South Korea already possess sizeable stockpiles, and Beijing has been relatively proactive in building more strategic petroleum reserve bases in recent years. Also, member states of the Association of Southeast Asian Nations (ASEAN) signed the ASEAN Petroleum Security Agreement (APSA) in March 2009 as a hedge to future fossil fuel supply

---

[8] For instance, see "World Oil Demand Growth to be Led by Asia—IEA", in: *Reuters* (10 November 2009).

disruption—also a sign that Asia-Pacific countries continue to take seriously the issue of energy supply security.

While fossil fuels will continue to make up the bulk of energy mixes in the Asia-Pacific, developing Asian countries are exploring other sources of energy to diversify their energy mixes and reduce their vulnerability and dependence on oil. However, these alternative energy options also have their pros and cons. The next section highlights some of the alternative energy options that are being considered in contemporary energy policymaking circles in the Asia-Pacific, largely focusing on the Northeast and Southeast Asian subregions.

1.4.1 Growing Emphasis on Clean Energy Investments

Asia has moved increasingly towards harnessing clean energy. China, for instance, which continues to seek new alternative sources of fossil fuel to meet its domestic demands, has also taken steps to bolster national clean energy R&D, as seen in the establishment, by China's National Energy Administration (NEA), of a network of 16 national energy centres in January 2010, focusing on clean energy in particular.[9] Renewables also play a significant part in China's far and remote territories, such as solar energy projects in Tibet,[10] which is China's richest resource of solar energy. Locations suitable for wind energy—aside from the eastern coast—include China's Xinjiang province in the northwest[11] and northern territories bordering Mongolia.[12] Besides China, other Asia-Pacific countries are also striving to harness clean energy in order to diversify their national energy mixes (Caballero-Anthony et al. 2010). According to energy experts examining the 'green race' in Asia, regional investments in the clean energy sector are believed to likely double in 2010 to nearly US$70 billion in total.[13] This is definitely a sign of the increasing interest being evinced in Asia-Pacific countries to harness clean energy sources—a prospect that will likely be spurred by rapid economic recovery in the region—and the tangible moves being made towards their use.

---

[9] This network of 16 centres is said to be the first batch of such institutions, implying the intention of Beijing to establish more centres in times to come; see "China Sets Up First Batch of 16 Energy Research Centres", in: *AsiaPulse News* (7 January 2010).

[10] "Another 10-Megawatt Solar Plant Built in Tibet", in: *People's Daily* (23 March 2011), at: http://english.peopledaily.com.cn/90001/98649/7329105.html (22 December 2011).

[11] "China's Biggest Wind Farm to Rise in Xinjiang", in: *Asian Power* (19 November 2010), at: http://asian-power.com/project/in-focus/china%E2%80%99s-biggest-wind-farm-rise-in-xinjiang (22 December 2011).

[12] "Growing Pains of China's Wind Power Industry", in: *China Daily* (28 May 2011), at: http://www.chinadaily.com.cn/china/2011-05/28/content_12598392.htm (22 December 2011).

[13] In 2009, global clean energy investments had hit US$146 billion of which Asia already accounted for a third; see "Investments in Asia's Clean Energy Sector Likely to Reach US$70 Billion", in: *Channel NewsAsia* (20 May 2010).

On the flip side, it should also be recognised that the construction of energy infrastructure for such projects could potentially lead to environmental and bio-diversity degradation. For instance, the Laos-Thailand Nam Theun 2 (NT2) hydroelectric power project funded by the World Bank in 2000 could have flooded approximately 450 km^2 of the Nakai Plateau, which has rich biodiversity (Barton et al. 2004, p. 447). Such an event would have adverse impacts on inhabitants who depend on the ecosystem in the affected area for their livelihoods. For instance, in the case of the NT2 project, about 4,500 inhabitants needed resettlement and another 40,000 expected to be affected by ensuing damage to the local fishery industry (Barton et al. 2004, p. 447).

Biofuels is another promising prospect although it has seen several setbacks in the course of its development. First-generation biofuels—made out of corn and soybean—were reportedly unable to significantly help reduce GHG emissions or improve energy security,[14] and posed a threat to food security given the competitive use of land for energy crops versus food crops. While second-generation biofuels—from jatropha—were able to make up for this by growing on less fertile and mar-ginal lands, they required five times more water than the former.[15] Indeed, second-generation biofuels may be best suited for regions that receive high levels of rainfall. In the same way, although third-generation biofuels from algae have demonstrated ability to produce more oil yield than earlier bio fuel generations, they are capital and energy intensive. In this respect, it would seem that while there are clean energy alternatives to fossil fuels, such options are not without certain costs.

1.4.2 Asia Going Nuclear?

Another option that until early 2011 was increasingly popular with developing countries was that of nuclear energy. According to the International Atomic Energy Agency (IAEA), nuclear energy emits the least amount of carbon vis-à-vis other energy sources.[16] Developing countries had also perceived nuclear energy to be an easy fix and a cheaper option compared to other renewable technologies that were still relatively expensive. Most ASEAN countries have already announced their plans or deliberations on the use of nuclear energy in their energy mixes albeit with objections from some sections of their societies. The main concern for many of those objecting nuclear energy is the lack of safety standards, environ-mental impact assessments and transparency in many developing Asian countries.

---

[14] "Biofuels: OECD Report Blasts biofuels as "Costly and Ineffective""", in: *Europe Agriculture* (28 July 2008).

[15] McKenna, Phil, "All Washed Up for Jatropha?", in: *Technology Review* (9 June 2009), at: http://www.technologyreview.com/energy/22766/ (22 December 2011).

[16] Rogner, Hans-Holger, *Nuclear Power and Climate Change*, Presentation, at: http://www.iaea.org/OurWork/ST/NE/Pess/assets/03-01708_Rognerspeech.pdf (22 December 2011).

The sentiment against nuclear energy has gained ground particularly in the wake of the earthquake and tsunami that hit Fukushima, Japan, in March 2011, which resulted in a nuclear crisis and caused widespread ripple effects across Japan's energy, industry, agriculture and tourism sectors. The Fukushima incident was a critical reality check for East Asian countries that were considering nuclear energy as a means of sustaining their energy needs and economic development. The Japanese experience demonstrated that dealing with the latent nuclear safety issues would require more than reliance on advancements in nuclear science and technology. While there was much faith in nuclear technology, there was less so in Japan's nuclear industry and the processes surrounding the checks and safety standards in nuclear power plants. That said, however, although governments in Asia may be laying low with regard to their nuclear energy plans and ambitions, the nuclear option is nowhere near being taken off the table in the region.

1.5 The Way Forward for Energy Security

In sum, the NTS challenges posed by the volatile energy prices, global financial crisis, increasing demand, slow transition to alternative fuels and the anticipated climate change due to persistent use of fossil fuels would be considerable and exposes the world to a potential 'energy tsunami' (Iida 2008).[17] While there is a need to meet rising energy demand in the short run, energy efficiency and the use of cleaner alternative fuel sources have to be promoted, notwithstanding the present drop in oil prices, as the long-term solution. From both standpoints, it becomes imperative that energy-related investments are bolstered. Therefore, a rethink along the lines of the roles of technology, civil societies, markets and governance might be necessary in order to ensure energy security.

1.5.1 Harnessing Technology for Energy Security

In today's era of globalisation, technology has become increasingly relevant and relied upon for solving almost every problem that is encountered by man, improving his way of life and even furthering national goals, such as socioeconomic development. It is in this respect that technology and energy security are closely related to each other. With global warming becoming a perennial existential threat to mankind, and in the face of increased energy demand but uncertain supply, technology would seem to be the solution (Chew 2008).

---

[17] Should current energy security issues persist, the state and its citizens face possible future increment of food prices, economic deterioration in the form of inflation and unemployment, poor living environment as a result of pollution-induced health hazards and threats to sociopolitical stability.

1. *Traditional fuel sources.* The common perception that fossil fuel supplies would be exhausted in a matter of decades has been challenged within scientific circles. Therefore, the issue at hand is not one of unavailability of supplies, but rather a problem of accessibility. However, access to untapped fuel sources in naturally inhospitable regions requires better exploration and extraction techniques. With traditional fuels projected to dominate the energy mix in the near future, their continued usage without causing further ozone layer damage would hinge significantly on the newly conceived carbon capture and storage (CCS) techniques, such as the one being experimented in Norway.[18]
2. *Alternative energy sources.* As highlighted above, although various alternative energy sources are available, these have associated advantages and disadvantages. In this regard, it is important that countries fully assess the alternative energy source best suited for a specific circumstance and ways in which these different alternative energy sources can be integrated into the national energy mix.
3. *Supply chains.* The energy supply chain is concerned with not only exploration and extraction but also the refinement and distribution processes. Technological solutions will be required for addressing concerns related to geographical limitations and security risks affecting the supply chain. For instance, a possible focus for R&D could be mitigating power cable transmission losses to enhance energy efficiency. Technology might also help to ease the transportation costs associated with energy products, such as liquefied natural gas, by reducing reliance on overland pipelines that are subject to geopolitical disruptions.
4. *Energy efficiency and conservation.* More attention should be devoted to curbing energy demand through technological solutions for households and industries that help to conserve energy, and thereby to sustain energy resources. For instance, the use of fuel cells in vehicles, such as the venture by Rolls Royce to develop aircraft fuel cell engines, could significantly help to manage the consumption of petrol and diesel, which are seeing increasing demand as a result of rising affluence and improved lifestyle.[19]

Tremendous amounts of R&D investments would be required, however, to realise the technological dream of ensuring energy security. It is a fact that no existing technology is capable of achieving the recently set target of realising GHG emissions lower than 450–550 parts per million (ppm) by 2050 (Birol 2007). What is more, R&D often entails not only long gestation periods and considerable costs but also uncertain outcomes. While capital expenditure by leading energy firms did increase sharply in nominal terms over the first half of the century's first

---

[18] Potential non-technological problems of CCS, such as leakage from capture, would need to be addressed before such techniques are economically viable for widespread adoption; see "CO_2 Capture Key but Still Not Answer", in: *Upstream* (30 September 2005).

[19] Further research is still needed in order to reduce the cost of fuel cells and develop better output so that such systems can be miniaturised for wider application; see "Fuel Cells Hold Hope of Clean Energy for the Future", in: *The Straits Times* (28 April 2007).

decade, sustained capital investments were adversely impacted during the global economic recession in 2008–2009 due to tighter credit being made available by bankers, lower profitability and reduced energy demand (IEA 2010, p. 10). Then again, notwithstanding cumulative energy infrastructure investments projected to amount to nearly US$20 trillion (in 2005 dollars) over 2005–2030 (Pascual 2008; Bochkarev and Austin 2007), there still exists lingering uncertainty over the exact cost of discovering and exploiting energy sources over the coming decades.[20] Furthermore, the impact on additional energy capacity generation from such increased investment spending is being blunted by rising costs, not to mention the effects of the credit crunch. Investments in 2005, for instance, were reportedly lower than that in 2000 and capacity additions due to planned upstream investment until 2010 were expected to boost global spare crude production only slightly— regulatory delays playing a role here as well (Pascual 2008, p. 6). Such details drive home the fact that while securing reliable and affordable energy depends largely on adequate investments in R&D, this will only be possible in an environment of good market governance.

1.5.2 Instituting Good Market Governance for Energy Security

Since the end of the Cold War, the state has no longer been perceived as the principal provider of public goods. Riding the wave of globalisation and drawing benefits from the free market principles of competition, the provision of energy-related public goods is increasingly being shifted to the private sector. However, historical and contemporary antecedents point to the drawbacks of relying solely on the markets for competitively priced energy supplies. The 1973 oil crisis, for instance, saw a public rush for consumer goods in Japan as a consequence of inadequate government intervention and blind reliance on market principles toward securing foreign oil supplies.

The roles of markets and governance are intricately linked in contemporary energy security. Importantly, in the event of failure in the practices of the free market, or at the other end of the spectrum, interventionist mechanisms, it is the ultimate end-user, i.e. the average citizen, who is most severely affected. With energy prices exacting a toll on the average citizen, populist calls for increased government action have grown louder over recent years. However, government intervention in the name of energy security remains open to potential abuse by political leadership for parochial ends. For instance, the military junta in Myanmar

---

[20] Investments totalling over US$11 trillion are required in the power sector; capital expenditure is expected to be US$4.3 trillion and US$3.9 trillion in the oil and gas sectors respectively. About half of all energy infrastructure investments will be in developing states, where demand and production are projected to increase the fastest.

is purported to have exploited energy revenues to fund grandiose projects and weapon acquisitions even as the population at large subsists in abject poverty.[21]

Relying on free market or interventionist governance mechanisms alone, therefore, cannot ensure energy security—a combined approach melding both mechanisms could be the solution. Embracing the basic tenets of transparency, accountability, flexibility and the principles of respect for competition inter alia, good market governance could be defined as the best set of all laws, regulations, processes and practices that affect the functioning of a regulatory framework and the market (Hancher et al. 2004). It could provide an environment conducive for investors and, in particular, the increasingly important small- and medium-sized enterprises (SMEs) that occupy specialist niches in innovative energy solutions.

The Japanese example during the oil crises of the 1970s demonstrates a viable approach in which both free market and interventionist governance approaches were weaved together to ensure energy security. During the second oil crisis in 1979, consumers in Japan did not panic and rush to buy consumer goods, as had happened in 1973. Similarly, in the immediate aftermath of the Iraqi invasion of Kuwait and the Coalition military build-up prior to Operation Desert Storm in 1990, Japanese energy investors in the Middle East remained optimistic despite an anticipated, but never materialised, oil crisis.[22] This could be attributed to the post-1973 policies adopted by the Japanese government, such as energy-saving measures and guidelines defining public–private relationship, with respect to investments in energy efficiency and alternative energy technologies (Weatherford et al. 1989, pp. 605–608). Japan's economic performance immediately after the 1979 oil crisis, when compared to its counterparts, could be indicative of the success of its policies. For instance, in 1974 Japan's real gross national product (GNP), after the first oil shock in 1973, was -0.5 %, compared to the UK's -2.0 % and Italy's 4.2 %. However, in 1981, two years after the second oil shock, Japan's real GNP was registered at 5.0 %, compared to the UK's -2.3 % and Italy's 3.7 % (Weatherford et al. 1989, p. 615).

That said, however, the story seems to have taken a turn vis-à-vis Japan's current economic performance amidst its nuclear energy crisis. While Japan was able to reduce its dependency on oil imports, its partial self-sufficiency apropos its reliance on nuclear energy has become a bit of a vulnerability. A contributing reason for this could be the governance structures instituted in Japan for civil

---

[21] The Human Rights Watch has alleged that lucrative oil deals made with energy-thirsty nations, such as China and Thailand, has emboldened the ruling military junta to ignore calls for improving human rights and democratisation; see "Activists Urge UN to Impose Energy Sanctions on Myanmar", in: *Oil Daily* (21 November 2007).

[22] The export of petrochemical plant technology to the Middle East registered a significant surge (48.9 %) between April–September 1990, notwithstanding the Iraqi invasion and subsequent allied military build-up, compared to the same period in 1989. Even so, industrial officials admitted that it would have been almost impossible to resume plant construction in the region, if the war expanded, due to physical security concerns; see "Japanese Industry Shocked, but Copes Coolly with War (Iraq-Kuwait Crisis, 1990)" in: *Japanese Economic Newswire* (17 January 1991).

nuclear energy, which hitherto came under the auspices of the Ministry for Economy, Trade and Industry (METI), and as such gave limited leverage to other relevant ministries and government agencies, such as the Ministry for the Environment. This arrangement has changed post-Fukushima, as the Japanese government mulls over implementing a more multisectoral and multiministerial body to integrate energy- and environment-related matters.

There is increasing worldwide interest in reinvigorating the governmental role, in conjunction with the market approach, for addressing energy security issues, and this is notwithstanding the failure to agree on numerical carbon emission limits during the recent G8 Summit. For instance, in February 2008, the institution of an independent body, amalgamating member states' national energy regulators and equipped with flexible powers, was championed within the European Commission.[23] Earlier in 2006, there were calls in Thailand for the emplacement of an independent power regulator prior to the privatisation of the Electrical Generating Authority of Thailand.[24] Similarly, the United States Federal Energy Regulatory Commission (FERC) adopted new guidelines in 2007 to enhance accountability among market operators and hence promote competition.[25] These exemplify initiatives that balance free market principles with interventionist governance approaches and highlight the growing recognition of the importance of good market governance in ensuring energy security.

Intergovernmental efforts to ensure improved energy cooperation are also being seen. At the ASEAN level, for instance, member states have collectively agreed in principle to unite in a common stand to address climate change and call for bolder and significant cuts in GHG emissions by developed states.[26] In fact, ASEAN is acutely aware of the impact of climate change, which member states have broadly recognised to be affecting various aspects of human life, including food security.[27] Probably the most concrete action taken to date by ASEAN towards addressing the environmental impact of energy use is the ASEAN Plan of Action for Energy Cooperation (APAEC) 2010–2015, which stresses upon the need to "enhance energy security and sustainability for the ASEAN region including health, safety

---

[23] "Legal Battle Brews over EC's Regulatory Agency Plan", in: *EU Energy*, 176 (8 February 2008).

[24] "Thailand: Democrats Criticize Government's Energy Policy", in: *Thai News Service* (11 August 2006).

[25] "Proposed Changes Would Rise Accountability of Market Operators", in: *Platts Commodity News* (21 June 2007).

[26] "ASEAN United in Fight versus Climate Change; Agrees to Adopt RP's Call for 'Deep, Early Emissions Cuts'", in: *Philippines News Agency* (16 June 2009).

[27] For instance, Heherson T. Alvarez, a Philippine official tasked by then President, Gloria Macapagal-Arroyo, to head the Philippine delegation to Bonn for the United Nations Framework Convention on Climate Change (UNFCCC) in June 2009 remarked that "In ASEAN, creeping climate change is a common occurrence and its impact is a rising destruction on whole communities and food systems. We must respond to protect the region and its extremely vulnerable population."; see Ibid.

and environment through the accelerated implementation of action plans".[28] The energy-climate change nexus was once again reaffirmed as a primary agenda for ASEAN energy cooperation during the 28th ASEAN Ministers on Energy Meeting (AMEM), which was held in Vietnam in July 2010.[29] Moreover, a recent development (at the time of writing of the chapter) has been the call for continued efforts towards achieving energy connectivity in ASEAN by 2015.[30] In addition to this, the establishment of the ASEAN Infrastructure Fund[31] serves to complement efforts to improve energy efficiency and regional connectivity.

At the global level, the Global Green Growth Institute, spearheaded by South Korea and founded in June 2010, seeks to enhance the capabilities of developing states in adopting green growth economic strategies.

1.5.3 Engaging Civil Societies in Energy Security

A variety of definitions exist for the term 'governance'. However, with respect to energy security, it might be fitting to use the definition coined by the Organisation for Economic Co-operation and Development (OECD), which states that governance denotes the use of political authority and exercise of control in a society in relation to the management of its resources for social and economic development (Weiss 2000, p. 797). This broad definition encompasses the role of public authorities in establishing the environment in which economic operators function and in determining the distribution of benefits as well as the nature of the relationship between the ruler and the ruled (Weiss 2000, p. 797). While successful governance has been linked to democratisation and stable socioeconomic development, there are also instances of skewed distribution of economic benefits and protracted hardships leading to wider dissatisfaction in some Asian nations (Soerjono 2008).

The concept of governance has since evolved into a broader concept that encompasses all actors—state and non-state—other than the public sector who

---

[28] The overarching theme of APAEC 2010–2015 and a vast majority of its projects concerns clean energy technology development; see ASEAN, "ASEAN Plan of Action for Energy Cooperation 2010–2015: Bringing Policies to Actions: Towards a Cleaner, More Efficient and Sustainable ASEAN Energy Community", at: http://www.aseansec.org/22675.pdf (25 October 2010).

[29] ASEAN, "Joint Media Statement of the 28th ASEAN Ministers on Energy Meeting (AMEM)", Da Lat, Vietnam, 23 July 2010, at: http://www.asean.org/24940.htm (25 October 2010).

[30] ASEAN, "Joint Ministerial Statement of the 29th ASEAN Ministers on Energy Meeting (AMEM)", Jerudong, Brunei Darussalam, 20 September 2011, at: http://www.asean.org/26626.htm (22 December 2011).

[31] "ASEAN Launches $650 m Fund to Build Infrastructure", in: AsiaOne (26 September 2011), at: http://www.asiaone.com/News/Latest%2BNews/Asia/Story/A1Story20110926-301458.html (22 December 2011).

could be involved in achieving sustainable human development objectives at the corporate, national, regional and global levels.[32] Among the non-state actors involved in governance, civil societies have gained increased prominence. For instance, in Indonesia, the civil society is increasingly involved in the process with occasional invitations to ad hoc government and public consultations. In the area of energy security, civil societies have served as advocates for environmental protection, acting as 'watchdogs' of government policies, with a role to play in raising and shaping public awareness on supporting or rejecting government initiatives (Alexandra 2008). Such efforts have been aided by advancements in sophisticated information and communications technology (ICT), the use of which has become increasingly pervasive in society. Many civil societies have even been able to conduct their own research and release reliable information that serves as an alternative to official government publications. Backed by sufficient data, they have been increasingly successful in pushing through their agenda especially when it comes to energy security issues. A case in point is the permission granted by a federal court to environmentalist groups, Friends of the Earth and Greenpeace, to proceed with a global warming lawsuit against two US government agencies that fund oil and gas projects.[33]

In the case of East Asia, however, civil societal involvement in governance is still at the nascent stage and there is room for improvement along the lines of: (a) creating a comprehensive agenda among civil societies in order to gain greater bargaining power with the government; (b) continuing to work through formal processes, and to demand more access and transparency from governments; and (c) strengthening civil society networks at both domestic and regional levels which could help in expanding its knowledge base and advocacy power (Alexandra 2008).

1.6 Conclusion

In the face of persistent energy supply and price volatilities as well as surging global demand, countries have traditionally viewed energy security as a matter of ensuring adequate, affordable and reliable energy supplies. However, the projected increase in consumption of and continued reliance on fossil fuels entail far-reaching environmental and socioeconomic consequences beyond the notion of mere supply security. These concerns include climate change, with its attendant problems of rising sea levels and risks posed to the ecosystem, as well as socio-political impacts in the face of public dissatisfaction over volatile energy prices.

---

[32] Non-state actors include the private sector as well as civil societies.

[33] Environmentalists alleged that the Overseas Private Investment Corporation (OPIC) and Export–Import Bank of the United States (Ex-Im) had illegally provided over US$32 billion in financing and insurance for energy projects over the past decade without assessing their impact on global warming as required under the National Environmental Policy Act; see "Global Warming Case against US to Proceed", in: *Oil Daily* (25 August 2005).

The present chapter demonstrates that contemporary energy security no longer concerns merely the security of supply but is also about related NTS issues, be it environmental impact or socioeconomic development. Even so, as seen from the energy security situation in many Asia-Pacific countries, while governments have come to recognise emergent energy-related NTS concerns, their short-term, interim goals when it comes to ensuring energy security primarily emphasise securing traditional energy resources to manage supply. The long-term energy security approach, meanwhile, would be one that includes the management of not only its supply but also demand.

Several roadblocks stand in the way of energy security in East Asia, particularly due to the lack of proper and adequate governance structures for regulating market forces and the implementation of prudent policies that could ensure sustainable socioeconomic development. Given rising energy demand, the continued heavy reliance on fossil fuels and the dire consequences of global warming, long-term solutions for this dilemma lie no doubt in ensuring energy efficiency and in the development of alternative fuel technologies, both of which call for extensive R&D initiatives. However, R&D investments, even if critical, entail long gestation periods and immense costs, and often have uncertain outcomes. What is more, lowered energy prices can be expected to reduce incentives for continued R&D in alternative energy sources while encouraging unrestrained consumption. Given such ground realities, the following recommendations are proposed:

1. *Harnessing technology.* Technological advances could lead to fossil fuels being utilised in ways that are more efficient and less polluting, thereby increasing environmental sustainability. Similarly, solutions that boost energy efficiency and the use of alternative fuels would not only curb the consumption of fossil fuels but also provide a diversity of cleaner energy sources that contribute to the long-term sustainability of energy supplies.

2. *Instituting good market governance.* Ensuring energy security is not solely the responsibility of either the market or the government. Adequate amounts of government intervention are required to mitigate market failures. Good market governance, which combines proper regulatory mechanisms and free market fundamentals, would engender consumer and investor confidence so as to foster an environment conducive for energy-related investments.

3. *Engaging civil societies.* While the state still retains a viable role in ensuring energy security, the importance of non-state actors, in particular civil societies, should not be discounted. More could be done to engage these groups which, with the aid of modern media communication tools and access to transnational networks, help to raise public awareness on concerns such as energy-related environmental and socioeconomic issues, monitor government policies as well as advocate alternative energy security solutions.

In conclusion, to engender support for and investments in the long-term energy security approach, more effort is required in strengthening the role of governance in concert with the free market. The role of non-state actors, such as civil societies, would become increasingly important especially in helping to ensure public accountability and good governance. To bolster energy investments, good market governance—not solely free market principles or an interventionist approach—is necessary to foster an environment conducive for investors and consumers to curb energy consumption, create innovative solutions to make fossil fuels more environment-friendly and expand the applicability of alternative energy sources in the near- and long-term future. This concerted approach by all actors—state and non-state—would help to eventually ensure sustainable human development through adequate access to a diversity of energy resources, and thus protect the environment.

References

Alexandra LA (2008) The role of Indonesian civil society in energy security. Paper for the regional workshop on energy and non-traditional security, Singapore, 28–29 Aug

Andrews-Speed P, Ma X (2008) Energy production and social marginalization in China. J Contemp China 17(55):247–272

Ardiansyah F (2008) Environmental perspective on energy development in Indonesia. Paper for the regional workshop on energy and non-traditional security, Singapore, 28–29 Aug

Barton B, Redgwell C, Ronne A, Zillman DN (eds) (2004) Energy security: managing risk in a dynamic legal and regulatory environment. Oxford University Press, Oxford

Birol F (2007) Energy security: investment or insecurity. In: Coping with crisis, CWC working paper series 14 May International Peace Institute

Bochkarev D, Austin G (2007) Energy sovereignty and security: restoring confidence in a cooperative international system. Policy Paper, 1:1–33 (East-West Institute)

Caballero-Anthony M, Punzalan K, Koh C (2010) Renewable energy: a survey of policies in East Asia. In: NTS alert, 2 (March). http://www.rsis.edu.sg/nts/HTML-Newsletter/alert/NTS-alert-mar-1002.html (Accessed 22 Dec 2011)

Chew A (2008) Technology developments in energy security. Paper for the regional workshop on energy and non-traditional security, Singapore, 28–29 Aug

EIA Energy Information Administration, 2008: International energy outlook 2008 (September) (Washington D.C.: EIA). http://www.eia.gov/forecasts/archive/ieo08/pdf/0484(2008).pdf. Accessed 22 Dec 2011

Hancher L, Larouche P, Lavrijssen S (2004) Principles of good market governance. In: Tijdschrift voor Economie en Management 49(2):339–374

Iida T (2008) Non-traditional dimensions of energy security: socio-economical impact in Northeast Asia. Paper for the regional workshop on energy and non-traditional security, Singapore, 28–29 Aug

IEA (International Energy Agency) (2002) World energy outlook 2002. OECD/IEA, Paris

IEA (International Energy Agency) (2007) World energy outlook 2007: fact sheet—global energy demand. OECD/IEA, Paris

IEA (International Energy Agency) (2008) Worldwide trends in energy use and efficiency: key insights from IEA indicator analysis. OECD/IEA, Paris

IEA (International Energy Agency) (2010) The impact of the financial and economic crisis on global energy investment. IEA background paper for the G8 energy ministers meeting. 24–25 May 2009, OECD/IEA. http://www.iea.org/ebc/files/impact.pdf. Accessed 4 Feb 2012

Jamil S, Gong L, Caballero-Anthony M (2010) Dependency and complacency. in the energy sector: implication for human security, NTS alert, p 1, Oct

Mendoza MNF (2008) The socioeconomic impact of energy security in Southeast Asia. Paper for the regional workshop on energy and non-traditional security, Singapore, 28–29 Aug

Pascual C (2008) The geopolitics of energy: from security to survival. Brookings Institution, Washington, pp 1–16

Soerjono AR (2008) Towards a new framework: states, regional and global governance. Paper for the regional workshop on energy and non-traditional security, Singapore, 28–29 Aug

UNDP (United Nations Development Programme) (1994) Human development report 1994—new dimensions of human security. Oxford University Press, New York

Weatherford MS, Fukui H (1989) Domestic adjustment to international shocks in Japan and the United States. Int'l Org 43(4):585–623

Weiss TG (2000) Governance, good governance and global governance: conceptual and actual challenges. Third World Q 21(5):795–814

Yergin D (2006) Ensuring energy security. Foreign Aff 85:2

Abbreviations

AMEM	ASEAN Ministers on Energy Meeting
APAEC	ASEAN Plan of Action for Energy Cooperation
APSA	ASEAN Petroleum Security Agreement
ASEAN	Association of Southeast Asian Nations
Btu	British thermal unit
CCS	Carbon capture and storage
CHS	UN Commission on Human Security
CO_2	Carbon dioxide
EIA	Energy Information Administration
FERC	Federal Energy Regulatory Commission, United States
GHG	Greenhouse gas
GNP	Gross national product
IAEA	International Atomic Energy Agency
ICT	Information and communications technology
IEA	International Energy Agency
IPCC	Intergovernmental Panel on Climate Change
ISEAS	Institute of Southeast Asian Studies
METI	Ministry for Economy, Trade and Industry, Japan
mmbd	Million barrels per day
NEA	National Energy Administration, China
NT2	Nam Theun 2 hydroelectric power project
NTS	Non-traditional security
NTU	Nanyang Technological University
OECD	Organisation for Economic Co-operation and Development
OPEC	Organization of the Petroleum Exporting Countries
ppm	Parts per million
R&D	Research and development
RSIS	S. Rajaratnam School of International Studies
SMEs	Small- and medium-sized enterprises
UNDP	United Nations Development Programme
UNFCCC	United Nations Framework Convention on Climate Change

Author Biographies

Mely Caballero-Anthony (Singapore): Associate professor at the S. Rajaratnam School of International Studies (RSIS), Nanyang Technological University (NTU), Singapore, and head of the RSIS Centre for Non-Traditional Security (NTS) Studies. She has also served as director of external relations in the Political-Security Community Department, Association of Southeast Asian Nations (ASEAN) Secretariat. Her research interests include regionalism and regional security in the Asia-Pacific, multilateral security cooperation, politics and international relations in ASEAN, conflict prevention and management as well as human security. Her latest publications, both single-authored and co-edited, include: *Political Change, Democratic Transitions and Security in Southeast Asia* (Abingdon: Routledge, 2010); *Understanding Non-Traditional Security in Asia: Dilemmas in Securitization* (UK: Ashgate, 2006); (with Acharya, A.; Emmers, R.): *Studying Non-Traditional Security in Asia: Trends and Issues* (Singapore: Marshall Cavendish, 2006); *Regional Security in Southeast Asia: Beyond the ASEAN Way* (Singapore: ISEAS, 2005); (with Acharya, A.): *UN Peace Operations and Asian Security* (Abingdon: Routledge, 2005).

Swee Lean Collin Koh (Singapore): Associate research fellow at the S. Rajaratnam School of International Studies (RSIS). He was a research analyst on energy security at the RSIS Centre for Non-Traditional Security (NTS) Studies from 2008 to 2010, during which time he contributed to the centre's in-house publications, primarily focusing on energy security. He is also co-editor of a book (with Basrur, R.): *Nuclear Power and Energy Security in Asia* (Routledge, forthcoming in May 2012).

Sofiah Jamil (Singapore): Adjunct research associate at the S. Rajaratnam School of International Studies (RSIS). Sofiah was previously associate research fellow at the RSIS Centre for Non-Traditional Security (NTS) Studies, where she co-led two programmes—Climate Change and Environmental Security, and Energy Security. She writes regularly for the centre, focusing on policy-relevant perspectives on the environment in the Asian region. She is keenly interested in the role of civil society in human security and environmental issues. Her recent publications include: "Islam & Environmentalism: Greening Our Youth" (*in Igniting Thought, Unleashing Youth: Perspectives on Muslim Youth and Activism in Singapore*, 2 edited by M Nawab and F Ali; Singapore: Select Books, 2009). She continues to contribute to the work of the RSIS Centre for NTS Studies while pursuing her PhD (International, Political and Strategic Studies) at The Australian National University.

Chapter 2
The Role of Indonesia's Civil Society in Energy Security

Lina A. Alexandra

Abstract Energy (in)security has re-emerged as one of the central security issues for countries around the world. Despite its huge and diverse reserves of fossil fuel energy, Indonesia finds itself among countries suffering from an energy crisis. This paper is concerned with the lack of a comprehensive response by the Indonesian government to the domestic energy crisis. Troubled by these circumstances, non-state actors (or civil society) in Indonesia have chosen to act in response to the government's "failure". The role of civil society in energy security will be elaborated upon, particularly how non-governmental organisations (NGOs) engaged in the Indonesian energy sector perceive and propose initiatives that target the government, public and other relevant stakeholders. The basic argument is that NGOs in the Indonesian energy sector assume key functions by, first, mapping out problems in current energy policies and, second, educating Indonesian society on the energy crisis and encouraging new ideas on creating an alternative energy supply and using this effectively.

Keywords Energy security · Indonesia · Non-governmental organisation · Civil society · Energy crisis · Electricity

2.1 Introduction

Energy (in)security has re-emerged as one of the central security issues for countries around the world. The available supply of energy is unable to meet the increasing demand (especially for fossil fuels) arising from having to feed

L. A. Alexandra (✉)
Department of Politics and International Relations,
Centre for Strategic and International Studies (CSIS),
The Jakarta Post Building, 3rd Floor, Jl. Palmerah Barat 142-143,
Jakarta 10270, Indonesia
e-mail: lina_alexandra@csis.or.id

M. Caballero-Anthony et al. (eds.), *Rethinking Energy Security in Asia:*   21
A Non-Traditional View of Human Security, SpringerBriefs in Environment, Security,
Development and Peace 2, DOI: 10.1007/978-3-642-29703-8_2, © The Author(s) 2012

industrialisation processes and sustain both economic growth as well as changing lifestyles. The total oil consumption worldwide in 2010 reached nearly 4.03 billion tonnes—a 3.1 % increase from 2009.[1]

In mid-2008, the price of oil soared to US$150 per barrel, triggering predictions that it might reach US$200 per barrel by the end of the year. Remaining relatively stable for a year or so, the price spiked again from late 2010 to mid-2011, eventually reaching US$116 per barrel due to uprisings in several oil-producing countries in the Middle East, such as Egypt and Libya.[2]

This situation has forced some countries to introduce unfavourable policies— such as reducing oil subsidies and enforcing energy efficiency—that could cause public protests and domestic instability. Indeed, several economies have fallen into ruin as increasing energy prices have set off a chain reaction of increasing prices for basic commodities. In the case of Indonesia, rising oil prices have pushed policymakers and observers alike to engage in a serious debate to reassess the country's policy on subsidising fuel.[3]

Despite its huge and diverse reserves of fossil fuel energy, Indonesia finds itself among countries suffering from the energy crisis. Based on research conducted by the Indonesian Institute for Energy Economics (IIEE) in 2006–2007, energy shortages have been occurring in almost all parts of the country (IEER 2007, pp. 8–13). Blackouts have been common in the capital, Jakarta, since at least as early as mid-2008, incurring large losses for the industrial sector. At least one headline has suggested that Japanese investors plan to withdraw from Indonesia should the energy crisis continue to disrupt production processes.[4]

The Indonesian government's plans to meet demand include the implementation of once-a-month weekend production days to compensate for two-hour reductions during weekdays. The government has also launched a national 10,000-megawatt (MW) coal-based power programme that is expected to be

---

[1] "BP Statistical Review of World Energy (June 2011)", at: http://www.bp.com/assets/bp_internet/ globalbp/globalbp_uk_english/reports_and_publications/statistical_energy_review_2011/STAGING/ local_assets/pdf/statistical_review_of_world_energy_full_report_2011.pdf (26 September 2011).

[2] Kollewe, Julia, "Oil Prices Rise Again on Further Libyan Fighting", in: *Guardian* (4 March 2011), at: http://www.guardian.co.uk/business/2011/mar/04/oil-prices-rise-again-libyan-fighting (26 September 2011).

[3] One example is the recent study released by the Department of Economics, Centre for Strategic and International Studies (CSIS), Indonesia. The study recommends that the government re-evaluate the subsidy policy. Specifically, it suggests increasing the price of the "premium" category, eventually easing the burden on the national budget. See Rangga D. Fadillah, "Late Fuel Subsidy Removal Hurts RI", in: *The Jakarta Post* (11 May 2011), at: http:// www.thejakartapost.com/news/2011/05/11/late-fuel-subsidy-removal-hurts-ri.html (26 September 2011).

[4] "Krisis Listrik Ancam Investasi" (Electricity Crisis Is Threatening Investments), in: *Kompas* (10 July 2008).

completed in 2011.[5] However, Indonesia's reserves of coal—the main component of the programme—are insufficient to meet foreign demand, thus painting an uncertain future for the plan.

Indonesia is highly dependent on fossil fuels, particularly oil, which is the main energy source for industry, transportation and even electricity generation. The country used to be a member of the Organization of the Petroleum Exporting Countries (OPEC), but its membership was suspended due to recent decreases in its oil production. In 2009, Indonesia produced 47.9 million tonnes of oil; in 2010, this went down slightly to 47.8 million tonnes. With domestic consumption rising from 59.2 million tonnes of oil in 2009 to 59.6 million tonnes in 2010, the country has needed to import oil to meet demand.[6]

Rising oil prices have thus compelled the government to gradually lift the national subsidy, in order to manage the country's budget. In 2011, the subsidy was around 8.2 % of total national expenditures (almost US$68 billion) in order to maintain a price of US$0.50 per litre for premium fuel (Aswicahyono et al. 2011). Meanwhile, a lack of adequate infrastructure means that the government remains unable to shift the balance of energy consumption towards alternative sources, such as coal and natural gas.

A central issue is the response by governments and related institutions to the energy crisis. On average, this has been minimal and not comprehensive enough to tackle the problem. Some believe that the government remains largely unaware of the seriousness of the crisis or how it might lead to further instability. During a public ceremony in May 2008, President Susilo Bambang Yudhoyono merely encouraged the public to be "optimistic", despite expectations of energy, food and economic crises.[7] While optimism might prove necessary, concrete action from the government will be far more crucial in addressing the needs of the people.

The negative impacts of the crisis on the people, combined with apparent government cluelessness when it comes to finding real solutions, has pushed Indonesia's civil society into playing a role. This paper thus focuses on the role of Indonesian civil society with regard to the country's overall energy security. In particular, it highlights how non-governmental organisations (NGOs) in the energy sector look at and respond to the issue of energy security in relation to the two primary stakeholders: the government and the public. This chapter poses several questions: how

---

[5] Alfian, "PLN Secures 65 % of Financing for First 10,000 MW Program", in: *The Jakarta Post* (31 January 2009), at: http://www.thejakartapost.com/news/2009/01/31/pln-secures-65-financing-first-10000-mw-program.html (26 September 2011); see also: "PLN Ready to Start 2nd 10,000 MW Program", in: *The Jakarta Post* (20 August 2009).

[6] "BP Statistical Review of World Energy (June 2011)", at: http://www.bp.com/assets/bp_internet/globalbp/globalbp_uk_english/reports_and_publications/statistical_energy_review_2011/STAGING/local_assets/pdf/statistical_review_of_world_energy_full_report_2011.pdf (26 September 2011).

[7] Simamora, Adianto P., "SBY: Keep Spirits Up Amid Looming Crises", in: *The Jakarta Post* (21 May 2008).

do NGOs perceive the problem of energy security in Indonesia? What are their responses, and to what extent can these responses be seen as successful?

I argue that NGOs in the Indonesian energy sector have played a crucial role. Most important, they have brought to attention the reality that current energy policies are ad hoc and incomprehensive, and thus need to be improved. The NGOs have also equipped society with a better understanding of national energy policies, offering guides on how to creatively develop alternative energy resources. Significantly, these organisations consider the government to be a part of the solution rather than a source of the problem, and as such have taken a "soft", more sophisticated approach that involves navigating formal mechanisms. In this way, these NGOs have learned to work through governmental channels rather than solely through informal processes.

This chapter is divided into three parts. The first describes the policies implemented by the government to deal with the energy crisis and the NGO response to them. The second analyses the various roles played by NGOs in the energy sector, illustrating the use of the soft approach. The third part elaborates on the challenges ahead in dealing with Indonesia's energy crisis, and explores the opportunities available to maximise the effectiveness of various solutions. Finally, recommendations are offered to highlight what may be done in the future.

2.2 Energy Security Policies and NGO Responses

The term "energy security" is not easy to define. The definition can differ in developed versus developing countries, as the reality of energy scarcity varies. For developed countries, energy security is defined as "reliability of supply, access to the energy resources in sufficient amounts, affordability, and protection from energy supply interruptions". In developing countries, the emphasis is placed "first and foremost [on] access to energy to supply basic needs like clean water, cooking, lighting, and public transportation" (Luft and Korin 2009, pp. 5–6). While Indonesia is considered a developing country, for the purposes of this chapter, which explores the role of civil society, it would be more useful to adopt the definition used for developed countries with its focus on energy supply and access—rather than extend the discussion to include the broader implications of energy issues.

Regardless of the definition used, the energy crisis in Indonesia remains framed in largely narrow terms, that is, with reference to supply issues. This can be seen in a particular piece of Indonesian legislation, Energy Law No. 30/2007. Chapter 6(1) of that law notes that an energy crisis is a condition in which energy (supply) is lacking, while Chapter 6(2) states that an energy emergency is a condition in which energy supply is disrupted as a result of the destruction of energy facilities and infrastructure.

There are at least three documents that reveal much about how the Indonesian government has responded to the current energy crisis: Presidential Regulation 5/2006, on the National Energy Policy; the National Energy Management

Blueprint (2006–2025); and the aforementioned Energy Law 30/2007. The regulations included in these documents clearly demonstrate that the National Energy Policy is focused on ensuring the security of the domestic energy supply (in particular, refer to the Presidential Regulation 5/2006, Chapter 2). The Policy is expected to achieve this in two ways: first, by achieving an "energy elasticity"[8] of one or less by 2025; and second, by creating a diversified mix of energy sources by 2025 to reduce national dependency on oil.

Moreover, the short-term policies adopted by the government also focus mainly on securing the country's energy supply. This involves optimising the production of energy and adjusting its price. In the National Energy Management Blueprint (2006–2025), this policy is translated into various programmes: rationalising oil prices by reducing both the fuel subsidy and the compensation (through direct cash assistance) given to the poor; shifting from kerosene to liquefied petroleum gas (LPG) for household use; increasing domestic oil exploration; developing new sources of alternative energy, including nuclear power plants; and building and strengthening Indonesia's energy-related infrastructure.

This supply-side approach largely ignores the problem of sufficient access to energy, that is, the demand side. While oil and gas supplies are abundant, not everyone enjoys equal access to them. Headlines in early 2008 trumpeted Jakarta's electricity crisis—as if it were a novel experience for Indonesia. In reality, however, most other parts of the country have experienced such a situation for years.[9] Enacted in response to the aims outlined in the previous paragraph, a 300 % increase in the price of kerosene has also created significant problems for low-income groups, as most do not have access to energy alternatives. In fact, the government's policy of shifting from kerosene to LPG has generated new problems, as LPG prices continue to rise while supplies remain uncertain (Indriyanto 2008).

The public regularly levels criticism at the fact that most Indonesians are unable to benefit from the country's plentiful fossil fuel reserves and renewable energy resources. Instead, easy access to energy is typically available only to mid- and upper-level income groups. While poverty has long been characterised as the main reason behind this situation, it remains clear that it is the government's responsibility to ensure that its people have sufficient access to the country's own energy resources. However, a centralised energy policy—dependent on central distribution, and without consideration for the development of local energy sources for local needs—has prevented a large section of Indonesian society from accessing

---

[8] Energy elasticity is the percentage change in energy consumption to achieve 1 % change in national gross domestic product (GDP).

[9] Indonesia's total power-plant capacity is 29,705 MW, 22,302 MW of which is meant for Java and Bali, and 7,403 MW for outside Java; however, peak electricity demand in Java alone has reached 17,000 MW. See Kelik Dewanto, "Proyek Pembangkit 10,000 MW Obat Krisis Listrik" (10,000 MW Electricity Generation Project Solution), in: Antara News (13 July 2008), at: http://www.antara.co.id/arc/2008/7/13/proyek-pembangkit-10000-mw-obat-krisis-listrik (26 September 2011). On the electricity crisis outside of Java, see Basri, Faisal, 2008: "Menjinakkan Krisis Listrik" (Taming the Electricity Crisis), in: Kompas (14 July 2008).

enough energy to sustain basic needs.[10] Moreover, the government has not seriously pursued the development of alternative energy sources such as bio fuels, despite the previously mentioned policy to establish a new energy mix by 2025.

The complexity of Indonesia's energy problems has forced the government to recognise that it needs to involve civil society in establishing a new national energy policy, from planning right up to implementation. This has translated into the establishment of partnerships between the government and NGOs to carry out research and development. Presidential Regulation No. 5/2006, Chapter 6, notes that the government may provide facilities and incentives for pioneering efforts in energy alternatives, clearly encouraging the participation of civil society in research. Moreover, Energy Law No. 30/2007, in Chapters 17 and 19, notes the importance of society's participation in creating both national and local energy plans, while also mentioning that energy development should be carried out with the public's interest in mind.

The government, especially the Ministry of Energy and the Parliament, has invited NGOs to share their views through public consultations, hearings and other meetings. On many occasions, members of Parliament have also received reports and recommendations submitted by NGOs. Most NGOs, however, continue to see government initiatives on energy security as lacking comprehensiveness. The initiatives are sectoral in nature, dealing separately with the security of energy supplies, the impact of the crisis on the economy, and the social aspect (the environment). In addition, initiatives elaborated upon in the National Energy Management Blueprint remain in the early stages, and lack detailed planning on implementation and financing. For these reasons, many expect that the Indonesian energy crisis will worsen in coming years.[11]

Most NGOs in the Indonesian energy sector essentially agree with the government's decision to reduce the fuel subsidy. This assistance has not helped those who were supposed to receive it, having instead mainly benefited the middle and upper classes. The subsidy has therefore indirectly encouraged an inefficient pattern of energy consumption that would ultimately harm Indonesia's energy supply position in spite of its reserves. Furthermore, the government had not carefully scrutinised the subsidy plan before implementation—for instance, which fuel price to increase relative to another.

The 300 % increase in the price of kerosene is an indication of such mishandling. The decision to phase out kerosene in favour of LPG for household use, officially implemented in mid-2007, seems to have been made solely as a short-term solution. There was no preliminary research on certain regions' potential to develop sources of renewable energy. Thus, despite the potential for biogas to be utilised in some regions, the conversion has been indiscriminately implemented throughout the country. Such a situation has triggered concerns over the possibility

---

[10] Personal interview with Kuki Soejachmoen from Yayasan Pelangi (September 2008).

[11] Personal interview with Mohammad Suhud from WWF Indonesia (July 2008).

of corruption, with some suspicions that certain parts of the government might have been involved with particular LPG suppliers.

In general, NGOs do admit that there are more opportunities today for the public to influence the National Energy Policy. Government officials, especially from the Ministry of Energy and Parliament, have taken to occasionally inviting NGOs for ad hoc public consultations and hearings when formulating new regulations, particularly during emergencies. In most cases, however, NGOs can express their concerns only through certain cooperative figures in the Parliament or the Ministry of Energy. Lobbying, rather than meeting formally with bureaucrats, has thus become the more effective way for NGOs to assert themselves and influence national energy policies.[12]

2.3 Types and Roles of NGOs in the Energy Sector

As the Indonesian people have directly experienced the impact of the country's energy crisis, many have begun researching the roots of the crisis, both domestically and internationally. New technologies, particularly the Internet, have allowed people far greater access to information, while simultaneously giving them the opportunity to express concerns and share knowledge on issues related to the energy crisis.[13] This phenomenon has shown Indonesian NGOs that the public may have very critical views on issues related to energy security.

The Indonesian public is generally free to voice its concerns and to propose initiatives to the government. However, in order to do so effectively, and influence policies, they would need to have access to balanced perspectives on an issue. This has not proved to be the case. An example would be the 2003–2004 discussions on the calculation of the new fuel subsidy. At that time, there was practically no information on the official websites of either the Indonesian Ministry of Energy or Parliament concerning how NGO reports and recommendations would be utilised, information that could have indicated the extent to which those ideas influenced policymaking. The lack of transparency is further reflected in the fact that no official reports exist of discussions of certain policy drafts (energy-related or otherwise) currently under parliamentary process.

The role of NGOs, then, is twofold. On the one hand, such organisations assist the government by providing inputs and even critiques that can help to shape sound policies with greater benefits for the people, and by enhancing transparency in the decision-making process. On the other hand, NGOs play a crucial role in educating the public on the issues at hand. They can convince citizens not to

---

[12] Ibid.

[13] See the following useful blog articles: "Energy Security: A Look at Other Fuel Sources" (18 October 2006), at: http://www.post1.net/lowem/entry/energy_security_a_look_at (26 September 2001); and Ardian, Faddy, "Problem of Indonesia Renewable Energy Development" (18 February 2009), at: http://www.kamase.org/?p=582 (9 November 2011).

blindly support unwise energy policies. At the same time, they can also advance the people's understanding of the rationales behind certain policies, thus motivating them to support the government's energy efficiency goals. Moreover, NGO projects tend to be inclusive in nature. For example, projects aimed at identifying potential energy resources at the local level involve the participation of the community. Through these efforts, NGOs encourage the public to play a more proactive role in maintaining or even enhancing their access to energy, while also using available energy resources more efficiently; this could help to shift the blame for the various energy supply shortcomings away from the government.

NGOs go about expressing concerns over government initiatives in at least four ways: (1) undertaking projects on public education/awareness, either in support of or opposition to official initiatives; (2) empowering the people to generate their own energy resources, such as creating micro-hydro projects; (3) taking on a watchdog role with regard to government policies; and (4) engaging in advocacy by assisting the government in amending certain energy policies.

The typology used for categorising Indonesia's environment-focused NGOs can be useful in understanding the NGOs that play a part in the country's energy sector. These NGOs can be divided into three types: research-oriented, lobbying and mediating (Lin 1998, pp. 14–20).

Research-oriented NGOs are interested in academic research, to increase knowledge but also to stimulate public debate. They tend to offer science-based policy advice, both to the government and the public in general, perhaps assisting the government to design policies aimed at increasing energy efficiency. These NGOs tend to disseminate their findings through scientific publications, conferences and workshops, the media, etc. They tend to be "soft" in nature, emphasising collaboration with, rather than resistance to, the government. Lobbying NGOs stand in stark contrast to the research-oriented ones. Critical and radical in nature, they prefer using mass media to draw the public's attention to their own initiatives while also putting pressure on the government. To some extent, these NGOs also engage in advocacy, particularly when attempting to lobby against certain government policies. Mediating NGOs are unique in that they emphasise the establishment of networks both domestically and internationally. Their main interest is in creating connections with and providing information to interest groups and individuals that share their concerns. These networks are considered crucial, offering an effective way to share information as well as to reach consensus in pursuit of certain objectives.

According to general observation and interviews, most NGOs involved in the Indonesian energy sector are research-oriented.[14] They use publications (both digital and print), campaigns and empowerment projects to disseminate and implement their research. Of course, it is sometimes difficult to draw a clear line

[14] Information has been gathered from personal interviews with colleagues in WWF Indonesia, the Indonesian Institute for Energy Economics (IIEE) and Yayasan Pelangi. The author would like to extend personal thanks to Muhammad Suhud, Asclepias Indriyanto and Kuki Soejachmoen for their willingness to share knowledge and information that has been very useful to this paper.

between research-oriented, lobbying and mediating NGOs. In carrying out an empowerment project for instance, a research-oriented NGO may begin to function as a mediating NGO when it has to collaborate with other organisations to implement the project.

Let us look at a few examples of research-oriented NGOs that focus on energy research. Indonesian research-oriented organisations, or networks of these NGOs, conduct research and subsequently release their own publications that either support or reject government initiatives. These publications, including press releases, aim to educate the public about domestic energy reserves: the life span of the reserves, current and future domestic energy consumption, the impact of current consumption patterns, and recommendations on ways to deal with the energy crisis. The Indonesian Institute for Energy Economics (IIEE) is a prime example. It publishes the *Indonesian Energy Economics Review (IEER)* which offers analyses relating to energy security (IEER 2006, pp. 48–49). Volume II of the *IEER*, published in 2007, discusses domestic energy supply and demand, as well as how the attempt to achieve energy security should go hand in hand with sustainable development. IIEE is currently creating an energy database that will be made available to the public.[15] Another group, the Institute for Essential Services Reform, actually a network of NGOs, has since 2001, provided policy suggestions on promoting transparency and accountability in the Indonesian energy sector.[16]

An additional, and more practical, role of NGOs in this context is the release of research reports on various energy alternatives. A collaborative effort involving several NGOs—comprising Sawit Watch, Kehati, the Social and Economic Research Institute (INRISE), the Bogor Agricultural Institute and Media Indonesia Group—has focused on technical innovation and, in 2006, it led to the release of a report on developing palm oil as a bio fuel in Indonesia.[17]

Likewise, Yayasan Lembaga Konsumen Indonesia (YLKI, the Indonesia Consumer Association), provides the public with the reasons for supporting an increase in the electricity tariff. A similar programme, the Power Switch Campaign, launched by WWF Indonesia, was active in Jakarta between 2003 and 2007 as part of a worldwide campaign on energy efficiency.[18] WWF Indonesia is also currently implementing empowerment programmes in certain areas of Kalimantan (Borneo) in collaboration with the local government, assisting communities in assessing their region's potential for various types of renewable energy. The local

---

[15] Personal interview with Indriyanto from IIEE (July 2008).

[16] Details on the work of the Institute for Essential Services Reform can be found at "Institute for Essential Services Reform", *Electricity Governance Initiative*, at: http://electricitygovernance.wri.org/node/130 (26 September 2011).

[17] See "Dutch Import of Biomass: Point of View of Producing Countries on the Sustainability of Biomass Exports updated 070502", *Both ENDS* (2002), at: www.bothends.info/project/project_info.php?id=41&scr=st (26 September 2011).

[18] Personal interview with Mohammad Suhud (July 2008).

government uses the resulting assessment to determine how to allocate funding for the development of micro-hydropower plants.[19]

Yayasan Pelangi, an independent research institute, is another good example of a research-oriented NGO. The institution's lack of human resources has led to its developing networks and collaborations with other NGOs, in particular, those that focus on field activities. The organisation focuses on assessing ways to enable local communities to access energy and at the same time reduce emissions as a result of energy consumption. Yayasan Pelangi has partnered with a foundation within the Institut Bisnis dan Ekonomi Kerakyatan (IBEKA, an economic and business institute in Indonesia) to develop micro-hydropower projects in Aceh and South Sulawesi; its members also train local communities to operate these facilities. In addition, Yayasan Pelangi collaborates with Yayasan Dian Desa, which specialises in educating locals to build energy-efficient wood-stoves and kitchens. Yayasan Pelangi's latest project assesses the potential for biogas production in the Tangerang regency, following a request from the local government.

Meanwhile, several organisations and coalitions can be categorised as lobbying NGOs as they play a watchdog role. This role can be played out through both formal and informal approaches, including submission of reports and policy recommendations, attendance in public hearings in Parliament, participation in focus group discussions and other meetings with the Ministry of Energy, etc. However, this process has not been widely reported upon, and the Parliament's website has not yet included inputs from civil society in its online reports of public hearings.[20] This indicates that there is no guarantee that reports and recommendations offered by civil society will be taken into account by the government. At the same time, NGOs are continuing in their efforts to engage with the formal process, in the belief that they will eventually capture the attention of the relevant authorities.

In recent years, several NGOs have been very active in working to amend policies. Between 2001 and 2004, a coalition called the Working Group on the Power Sector approached Parliament and officials in the Ministry of Energy to propose an amendment to the Electricity Act. This coalition also provided information to Perusahaan Listrik Negara (PLN, the National Electricity Company) concerning the privatisation of the electricity sector, in order to equip the PLN to appeal to the Constitutional Court for a cancellation of the new Electricity Act, a move that was eventually successful.[21]

Another coalition has been pushing for an amendment to Undang–Undang Nomor 22 Tahun 2001 Tentang Minyak dan Gas Bumi (the Oil and Natural Gas Law Number 22 Year 2001, commonly known as UU Migas): this law is seen by many as giving too much leverage to private energy companies (Hutapea 2006). A more radical example is the activities of another NGO coalition, comprising Jaringan Advokasi Tambang (JATAM, the Mining Advocacy Network), Friends of

---

[19] Ibid.

[20] Personal interview with Indriyanto (July 2008).

[21] Personal Interview with Mohammad Suhud (July 2008).

the Earth Indonesia (also known as Wahana Lingkungan Hidup Indonesia [WALHI]), the Indonesian Center for Environmental Law (ICEL) and WWF Indonesia. This group has taken legal action against the mining activities of 13 companies operating in protected areas, previously justified under a presidential decree.[22] Finally, a coalition of NGOs, including Greenpeace, WALHI and the Indonesian Anti-Nuclear Society (MANUSIA), with support from local communities, has launched protests against a project to develop a nuclear power plant as an alternative energy source, as discussed in greater detail below.[23]

2.4 NGO Successes and the Soft Approach

Campaigns by Indonesian NGOs have had some notable results. One of the most significant was the halting of a nuclear power project that had been slated for construction at Mt Muria, on the north coast of Java. After massive public protests, the government agreed to reassess the plant's location. However, it is important to note that the government could still claim that civil society has failed to take into account the fact that nuclear power is cleaner than oil and an alternative to fossil fuels.

Other NGO successes include actions over the Electricity Act and UU Migas, as noted in the previous section. The Constitutional Court decided to cancel the Electricity Act after four years of struggle, during which NGOs heavily criticised the Act for legitimising unfair business competition.[24] The case of UU Migas can also be considered a success, albeit only a partial one. In 2003, NGOs forced the Constitutional Court to review the law, eventually judged to be against Chapter 33(2) and (3) of the Constitution. These provisions obligate the state to control sectors that are vital to the state and the people's basic needs; they also specify that only the state exercises ultimate control over the country's natural resources, which should be utilised for the welfare of the people. In 2004, however, the court decided that only three chapters of UU Migas were contrary to the Constitution, thus resulting in only a partial victory for the NGOs.[25]

Energy-saving campaigns by civil society have also achieved some success. A poll conducted by the newspaper *Kompas* in early July 2005 indicated significant public commitment to save energy. Around 71.7 % said that they had reduced

---

[22] "NGO Coalitions Will Take Legal Action on Indonesian Parliament's Endorsement of Government Decree", in *MAC: Mines and Communities* (24 July 2004), at: http://www.minesand communities.org/article.php?a=585 (26 September 2011).

[23] "A Nuclear Future?", in: *Down To Earth*, No. 72 (March 2007), at: http://dte.gn.apc.org/ 72nuk.htm (26 September 2011).

[24] "MK: UU Ketenagalistrikan Bertentangan Dengan UUD 45" (The Constitutional Court: The Electricity Bill Is against the Constitution), in *Tempo* (15 December 2004), at: http:// www.tempo.co.id/hg/nasional/2004/12/15/brk,20041215-34,id.html (26 September 2011).

[25] "Penyempurnaan UU Migas No. 22/2001 yang Berpihak kepada Kepentingan Rakyat".

their use of electricity, including for lighting purposes, while 34.4 % indicated that they had minimised the use of LPG for cooking.[26]

Several NGOs feel that the enactment of Energy Law No. 30/2007 was a relative success because it eventually gave details on the future energy mix. (The government had previously refused to do so, though it is unclear why.[27]) But other NGOs, such as IIEE, consider the law a failure, as plans for achieving sustainable development and energy security remain vague.[28] The law still lacks a detailed strategy, including a timetable for the achievement of targets set by the National Energy Plan, or an explanation as to how it is possible to achieve the desired energy mix even as total energy consumption is likely to increase annually.[29]

Beyond these quantifiable successes, another meaningful change is the shift in NGOs' ways of thinking. As referenced throughout this paper, recent years have seen many of these organisations take on a soft approach, focusing more on research and networking and less on confrontation with the government. Although subtler, this approach is generally found to be more effective, opening up possibilities of collaboration with the government in dealing with the Indonesian energy crisis.

How does this work? First, NGOs have learned to stop arbitrarily dismissing government initiatives without sound explanations. They have learned to provide sufficient data to support their arguments, information that can be useful for policymakers as well as the public. NGOs today seem to understand the importance of partnering with the government to create comprehensive and effective policies to deal with the energy crisis. As has been seen in the past, an overemphasis by NGOs on "harder" approaches could lead to the government being pushed aside when it should be the lead actor. Sound data can also equip NGOs to adopt a balanced approach when assessing government policies. This could lead them to support worthwhile policies despite their unpopularity with the public, thus creating a positive government–NGO relationship.

Second, NGOs have learned to utilise formal procedures, rather than merely working from outside governmental channels. In so doing, civil society is increasingly participating in the establishment of good governance in the energy sector, enabling NGOs to push the government to be more transparent in policymaking.

These achievements are important because the energy crisis in Indonesia is too large for the government or the NGO sector to face alone. Indeed, cooperation from all components of society will be necessary to deal successfully with the situation currently facing Indonesia.

---

[26] "Jajak Pendapat Kompas: Instruksi Penghematan Didukung dan Diragukan" (Kompas Polling: Efficiency Instruction Has Been Supported and Doubted), in: *Kompas* (18 July 2005).

[27] Personal interview with Asclepias Rachmi S. Indriyanto (July 2008).

[28] Ibid.

[29] Personal interview with Kuki Soejachmoen (September 2008).

2.5 Challenges

Despite the several successes mentioned above, challenges remain as civil society takes on broader roles in the Indonesian energy sector. The first difficulty comes from the governmental process. At public hearings in Parliament, NGOs are allocated very limited time to raise issues and explain their views. During such hearings, participating NGOs are typically consolidated into a single large group, resulting in each NGO being given only a minute or two to deliver its presentation. Also, often, the exercise seems like window-dressing—NGO representatives have on many occasions been asked to leave the hearing after their presentations, without having been given the opportunity to engage in further discussions.[30] In addition, to date, there is no mechanism for NGOs to find out if their recommendations are being considered.

At times, the government has also misused NGO support. For instance, the backing of civil society for a reduction of the fuel subsidy had the unintentional effect of allowing the government to massively increase the price of kerosene, despite this being the main fuel source for lower income groups.[31] One solution to this, considered effective by many, has been to engage with those government officials and members of Parliament who are cooperative, that is, those who are known to welcome and provide ample room for the views of NGOs in policy-making. Clearly it is important to maintain good communication with those politicians, as they offer entry points through which NGOs can influence the decision-making process and enhance transparency.

The second challenge comes from NGOs themselves, especially those not focused on the energy sector. In some cases, civil society groups involved in energy issues publicly support unfavourable government policies (for example, the decision to reduce the fuel subsidy), while in other cases they are quite critical. NGOs in other sectors perceive these actions as inconsistent. As a result, they challenge the actions of the energy-sector NGOs. The lack of a comprehensive agenda among NGOs on energy security issues proves particularly challenging when what is required is a united front to lobby the government on certain issues.

The third challenge arises from the existence of so-called "red plate" NGOs, which are established by the government or private companies to support government policies or special interests. Ostensibly on behalf of the public interest, these groups will, for instance, be very critical of a government decision to raise the price of a certain fuel (diesel, perhaps), while keeping silent when prices of other fuels are similarly increased.[32] Although easily seen through, this phenomenon poses a significant challenge to the integrity of other NGOs in the eyes of the public.

---

[30] Personal interview with Muhammad Suhud and Indriyanto (July 2008).

[31] Personal interview with Kuki Soejachmoen (September 2008).

[32] Ibid.

Similarly, a recent case known as the "blue energy" hoax might have ended up lowering public trust in new ideas and concepts that NGOs may wish to promote in the future. In late 2007, a man named Joko Suprapto claimed to have discovered an alternative fuel, known as blue energy, which purportedly could be processed from seawater. This alternative fuel was presented at the United Nations Climate Change Conference hosted by Indonesia in December 2007 in Nusa Dua, Bali. Soon thereafter, however, it was found that blue energy was unsupported by any scientific evidence. The Indonesian president, who had welcomed Joko Suprapto to the National Palace in appreciation of his work, suffered severe public criticism for having failed to thoroughly investigate the claim.[33] The worry now is that such incidents could sour the public's view of new—and potentially important—energy innovations.

The fourth challenge is the fact that only a limited segment of Indonesian society, in terms of both quantity and quality, can be reached through campaigns and public education efforts. This leads to an uneven level of knowledge about the energy crisis among the public, and it is clear that NGOs need to work harder to widen their reach. In fact, NGOs can increasingly take advantage of advanced information technologies to promote their activities and gain support for their campaigns. One small example of how this is already being done can be observed in a WWF Indonesia initiative. Its activists used methods such as sending short text messages (SMSes) by cell phone to promote their Earth Hour programme, which asks the public to switch off household electricity for an hour. Similar innovative use of technologies could significantly increase the reach of other such initiatives.

2.6 Conclusion

Indonesia, a country blessed with vast energy resources, faces a serious challenge from its ongoing energy crisis. One fundamental reason for this is a lack of comprehensive government planning on managing the country's valuable resources to supply domestic demand; currently the country simply focuses on export. The country's civil society plays a vital role in this situation: it needs to put pressure on and provide recommendations to the government, as well as educate the public on the challenges that Indonesia faces. Civil society also needs to be actively engaged in exploring how solutions may be found.

Although NGOs could voice their protest in a way similar to street demonstrations, one interesting discovery is their new preference for a softer and more sophisticated approach. Importantly, this can turn the government into a partner or part of the solution, rather than merely as a source of the problem. Using this

---

[33] Abdul Khalik, "SBY Faces Questions over 'Blue Energy' Hoax", in: *The Jakarta Post* (30 May 2008).

approach, NGOs have begun to engage with and struggle through formal mechanisms—such processes are not easy as many government officials remain reluctant to welcome this type of engagement. NGOs have also put more effort into developing recommendations for the government based on their own research, while several have released journals to circulate their research findings. On a more practical note, NGOs are also developing the capacity of local communities to fulfil their own energy needs by, for instance, developing micro-hydropower to generate electricity or campaigning for increased energy efficiency.

While these initiatives are certainly on the right track, Indonesian NGOs need to do more to engage the government and public as the main stakeholders in the country's energy crisis. Greater consolidation among the members of Indonesian civil society is clearly the agenda to pursue—to gain greater strength and, ultimately, to be more effective in influencing government policies for the benefit of the Indonesian people.

References

Aswicahyono H, Friawan D, Kartika P, Mugijayani W (2011) Penyesuaian Subsidi BBM: Pilihan Rasional Penyelamatan Ekonomi. (Adjusting the Fuel Subsidy: rational choice for saving the economy), In: CSIS policy brief (May)

Lin G (1998) Energy development and environmental NGOs: the Asian perspective. In: Cicero Working Paper, 3:5–33

Hutapea R (2006) Penyempurnaan UU Migas No. 22/2001 yang berpihak kepada kepentingan rakyat. Paper presented at national seminar and capturing session. Jakarta (14 Dec 2006), http://www.jatam.org/index2.php?option=com_content&do_pdf=1&id=109 (Accessed 26 Sept 2011)

IIEE (Indonesian Institute for Energy Economics) (2006) Surviving energy challenges. In: Indonesian energy economics review (IEER), 2:35–50 (Jakarta)

IIEE (Indonesian Institute for Energy Economics) (2007) Contesting energy security. In: Indonesian energy economics review (IEER), 2:35–45 (Jakarta)

Indriyanto AR (2008) Kerosene substitution programs in Indonesia: institutions and public policies. Presentation for the 31st IAEE pre-conference on clean cooking fuels and technology. Istanbul (16–17 June 2008). http://www.saee.ethz.ch/events/cleancooking/Ascelpias_Presentation_16Jun08.ppt (Accessed 26 Sept 2011)

Luft G, Korin A (2009) Energy security challenges for the 21st century: a reference book (Santa Barbara, California: Praeger Security International)

Abbreviations

CSIS	Centre for Strategic and International Studies (Jakarta)
GDP	Gross domestic product
IBEKA	Institut Bisnis dan Ekonomi Kerakyatan
IEER	*Indonesian Energy Economics Review*
IIEE	Indonesian Institute for Energy Economics
INRISE	Social and Economic Research Institute
LPG	Liquefied petroleum gas

MANUSIA	Indonesian Anti-Nuclear Society
NGO	Non-governmental organisation
OPEC	Organization of the Petroleum exporting Countries
PLN	National Electricity Company of Indonesia (Perusahaan Listrik Negara)
SMS	Short message service
UU Migas	Oil and Natural Gas (Law Number 22 Year 2001) (Undang–Undang Nomor 22 Tahun 2001 Tentang Minyak dan Gas Bumi)
WALHI	Indonesian Forum for Environment (Wahana Lingkungan Hidup Indonesia)
WWF	World Wildlife Fund
YLKI	Indonesia Consumer Association (Yayasan Lembaga Konsumen Indonesia)

Author Biography

Lina A. Alexandra (Indonesia) has been a researcher in the Department of International Relations, Centre for Strategic and International Studies (CSIS), Jakarta, since 2002. She is also a guest lecturer in the Postgraduate School of Diplomacy, Paramadina University, Jakarta. She is a contributor to *ASEAN's Quest for a Full-fledged Community* (CSIS 2007), and a member of the project team for "Nuclear Safety in Southeast Asia: Issues, Challenges, and Regional Strategy", conducted in 2008–09 with a report published in 2010. She will soon publish an article, co-written with M. Caballero-Anthony and Kevin D.G. Punzalan, titled "Civil Society Organisations and the Politics of Nuclear Energy in Southeast Asia: Exploring Processes of Engagement", in the edited volume *Nuclear Power and Energy Security in Asia* (Routledge).

Chapter 3
The Non-Traditional Security Perspective on Energy Security Policies in Singapore

Youngho Chang and Nur Azha Putra

Abstract The Singapore government treats energy security as a means of achieving sustainable economic growth. The state's 2007 National Energy Policy Report (NEPR) underlines the belief that an efficient energy market would drive economic growth and ensure the security of the state and its people. However, seen through a non-traditional security (NTS) lens, energy security must also encompass human security, that is, the welfare and development of individuals, households and communities. Building upon the NTS discourse, this chapter unpacks Singapore's energy policies and explores their implications for human security, focusing in particular on the challenges faced by the country's urban population.

Keywords Energy security · Non-traditional security · Human security · Singapore · Energy market

Y. Chang (✉)
HSS-04-65, Division of Economics, School of Humanities and Social Sciences,
Nanyang Technological University (NTU), 14 Nanyang Drive,
Singapore 637332, Singapore
e-mail: isyhchang@ntu.edu.sg
http://www.hssapps.ntu.edu.sg/faculty/econ.asp?u=isyhchang

N. Azha Putra
Energy Studies Institute, 29 Heng Mui Keng Terrace, Block A, #10-01,
Singapore 119620, Singapore
e-mail: azha@nus.edu.sg
http://www.esi.nus.edu.sg/our-researchers/nur-azha-putra

M. Caballero-Anthony et al. (eds.), *Rethinking Energy Security in Asia:*
A Non-Traditional View of Human Security, SpringerBriefs in Environment, Security,
Development and Peace 2, DOI: 10.1007/978-3-642-29703-8_3, © The Author(s) 2012

3.1 Introduction

No single, overarching definition of energy security exists. Energy security holds different meanings for different people at different times and at different locations. A nation's level of economic development also influences the way energy security is defined. Nevertheless, it is possible to identify certain basic ideas and concerns.

At its core, energy security, according to the International Energy Agency (IEA), refers to the "uninterrupted physical availability (of energy sources) at a price which is affordable, while respecting environment concerns".[1] Clearly, this definition leans heavily towards the notion of sustainable development, placing as it does a premium on economic growth and environmental sustainability.

Singapore's approach to energy security has been shaped by its lack of natural resources. The nation relies on oil imports for its refinery and petrochemical industries as well as its transportation sector, and on gas imports to generate electricity for its industries and households. The refining and trading of oil and the manufacturing of oil derivatives are key to the country's strategy for economic growth. This strategy has proven to be extremely successful. It is not surprising then that, as far as the government is concerned, its energy security objective is "to ensure stable, diverse and sustainable supplies of energy at competitive prices" (Ong 2009, p. 1).

Such definitions of energy security could, and should, be widened, particularly in view of the growing recognition that human security is a significant area of concern in the world today. A United Nations (UN) report in 2010 goes so far as to state that the principles of human security should be universalised. Accordingly, this chapter ponders the question of how Singapore's energy policies could be enhanced to increase human security. It does this by using the non-traditional security (NTS) framework to unpack the rationale behind Singapore's energy policies. The NTS framework uses human collectivities (and not the state) as the primary unit of analysis, and thus has useful conceptual tools beyond that available under a traditional state-centric framework.

With the NTS and human security discourse as its basis, this chapter explores the role of the energy industry in Singapore's economic development, and discusses the National Energy Policy Report (NEPR) of 2007 and the Economic Strategies Committee (ESC) report of 2010. This chapter also suggests that energy security can be achieved at the individual or household level by empowering consumers through a participatory process supported by an appropriate regulatory framework.

---

[1] See IEA (International Energy Agency), 2011, "Energy Security", at: http://www.iea.org/subjectqueries/keyresult.asp?KEYWORD_ID=4103 (17 May 2010).

3.2 The NTS Perspective on Energy Security

Traditional notions of energy security focus on the security of supply chains, which include extraction, processing, transportation and distribution. One of the energy security concerns in the early nineteenth century was the effect any shortage in oil supply could have on military preparedness. The military—the British Royal Navy being a case in point—had made the transition from coal-powered machines to ones using oil. With their rising dependency on oil and the geopolitical uncertainty surrounding the major oil-producing countries, oil-importing countries realised that their national security, and in particular their economies, was highly vulnerable to oil supply disruptions. Such security thinking has led to a preference for stratagems aimed at ensuring the security of supply chains, such as establishing petroleum reserves, diversifying oil sources and na-tionalising natural resources. Thus, a typical framing of energy security would be primarily concerned with three main issues: supply, distribution and access, that is, the ability of the state to procure energy supplies at an affordable price for its domestic consumption at a rate that is not detrimental to the nation's security and economic growth.

However, today, oil is no longer mainly consumed by the military. In fact, oil, coal and gas have been the most widely used fuels throughout the twentieth century and will continue to be so. They will account for nearly 80 % of total energy consumption by 2040 (ExxonMobil 2012, p. 1). In most nations today, oil is the predominant source of energy for the production of goods and services. It is commonly used in the transportation industry, in petrochemical products and as fuel for power plants. The use of oil by-products is so prevalent in public and private lives that an increase in the price of oil inevitably results in higher energy costs and therefore a rise in the price of other commodities, which will then affect national economies and household expenditures. Even in advanced economies, hikes in oil prices could reduce living standards dramatically. The poor, according to a 2007 UN Development Programme (UNDP) report on Asia and the Pacific, will be the worst affected:

> On the whole, rising oil prices have left the poor with few choices other than to cut back on their consumption of oil products or make other cuts in their household budgets. The urban poor tend to be worse off since they do not have the alternative of collecting fuelwood or other biomass.

It is against such a complex of vulnerability that the NTS perspective gradually evolved. In the post-Cold War era, the view that the focus of security should be widened beyond the state to include other referent objects such as human collectivities gradually gained traction. Emmers and Caballero-Anthony (2006, p. xiv) note that this was in part due to the growing recognition that security threats to the state are increasingly non-military in nature, and more the result of internal conflicts than wars. They outlined the defining features of the NTS approach: it focuses on non-military challenges to security; it recognises that most such challenges are transnational in terms of their origins, conception and effects; and finally, it moves beyond the state and regards human collectivities as the primary referent object of security.

Therefore, an NTS perspective broadens the energy security discourse with questions such as whether traditional strategies could ensure that all consumers have equal access to oil and gas-related goods and services such as petrochemical products, fuel and electricity, given that these have been made readily available for public and private consumption. After all, the availability of energy-related products at the individual, household and community levels is important but not necessarily sufficient for human development and welfare. There is still the issue of affordability, understood as the freedom to consume those products at a price that is not detrimental to human development and welfare.

Hence, of relevance to the NTS discourse is the notion of human security. At its barest form, human security is about human welfare and dignity. References to human security in contemporary discourse commonly attribute the concept to ul Haq. He is credited with universalising the notion through this evocative appeal:

> Human security is not a concern with weapons. It is a concern with human dignity. In the last analysis, it is a child who did not die, a disease that did not spread, an ethnic tension that did not explode, a dissident who was not silenced, and a human spirit that was not crushed. (ul Haq 1995, p. 116)

The notion of human security was formally adopted with the release of the UNDP's (1994) *Human Development Report*. The report states that human security has two aspects: "freedom from want" and "freedom from fear". Freedom from want refers to human safety from chronic threats such as hunger, disease and repression. Freedom from fear refers to protection from sudden and hurtful disruptions in the patterns of daily life at the individual, community and national level. The report lists seven categories against which threats to human security could be measured. They are economic security, food security, health security, environmental security, personal security, community security and political security.

The human security concept was further developed and articulated to the UN General Assembly in March 2010, when UN Secretary-General Ban Ki-moon presented his report on human security. Of significance is the addition of a third principle, "freedom to live in dignity", to the two principles mentioned earlier (UN 2010, p. 2). This principle suggests that any attempts to realise human security should include "mobilising communities through participatory processes" and "people-centred approaches" (UN 2010, p. 3).

> First, human security is in response to current and emerging threats … Second, human security calls for an expanded understanding of security where the protection and empowerment of people form the basis and the purpose of security. (UN 2010, p. 6)

Central to this notion of human security is the "instrumental role" of the government and the people. The report suggests that the government and the people have to forge a symbiotic relationship to address "the root cause of their weaknesses" and that the human security concept could help to "develop timely, targeted and effective responses that improve the resilience of Governments and people alike" (UN 2010, p. 4).

This implies that states need to work closely with their primary stakeholders, that is, their people, to ensure that security exists at the national and individual levels, and that, to a large extent, the security of states is interdependent with that of the people. After all, "no country can enjoy development without security, security without development" (UN 2010, p. 6).

3.3 Singapore's Present Energy Consumption Landscape

Singapore is an island city with limited energy resources. Currently, its energy market is reliant on oil and natural gas imports. In 2010, Singapore's electricity was generated by natural gas (78.7 %), petroleum products such as fuel oil and diesel (18.7 %) and renewables such as biogas, municipal solid waste and solar energy (2.6 %) (EMA 2011, p. 14). Singapore's energy mix is one of the least diversified in Southeast Asia (see Table 3.1 for a snapshot of the energy mix of the Southeast Asian states).

The IEA lists Singapore as the tenth largest importer of crude oil and oil products in the world in its 2011 *Key World Energy Statistics* report. According to International Enterprise Singapore (IE Singapore), about 82 % of Singapore's crude oil comes from the Middle East; almost 33 % of its crude oil is imported from Saudi Arabia. From within Southeast Asia, Vietnam is the largest supplier at 4.4 % (MTI 2007, p. 15).[2]

Gas is imported from Indonesia and Malaysia (MTI 2006). Since the state is almost wholly reliant on natural gas for its electricity, its dependence on just these two sources of supply renders it particularly vulnerable to supply disruptions and political risks.

Looking ahead, the government has decided to add liquefied natural gas (LNG) to the national fuel mix with the intention of diversifying its gas sources to meet future energy demand. An LNG terminal, which will enable it to import LNG from nations further away, is currently under construction (MTI 2006). According to the Singapore LNG Corporation, the terminal will cost approximately US$1.2 billion.[3] The terminal, which is expected to begin operations in 2013, will have the capacity to handle 3.5 million tonnes per annum. In 2010, the government announced that it

---

[2] Singapore imports its crude oil from Saudi Arabia, 32.8 %; Kuwait, 18 %; Qatar, 13.5 %; United Arab Emirates, 10.5 %; Other Middle East countries, 7.1 %; Vietnam, 4.4 %; Australia, 4.5 %; Malaysia, 3.7 %; and others, 5.4 %.

[3] Singapore LNG Corporation, 2010, "About the Singapore LNG Terminal", at: http://www.slng.com.sg/about-us-lng-terminal.html (16 January 2012).

Table 3.1 Association of Southeast Asian Nations (ASEAN) primary energy mix, 1995 and 2007

Fuel type	1995 (%)	2007 (%)
Gas	16.4	21.4
Coal	4.6	14.8
Oil	43.6	36.2
Hydro	1.0	1.2
Geothermal	2.1	2.9
Others	32.3	23.5
Total	100.0	100.0

Source Adapted from IEEJ/ACE (2011, p. 8); based on figures from the International Energy Agency (IEA) and the Lao PDR Ministry of Energy and Mines

will build a third LNG tank, which would raise the terminal's storage capacity to six million tonnes per annum.[4]

3.4 Unpacking Singapore's Approach to Energy Security: From Industrialisation to Sustainable Economic Growth

Energy security policies in Singapore are premised on the belief that energy resources are the means to achieving and sustaining economic growth. To date, the two most comprehensive documents that inform the state's approach to ensuring energy security are the NEPR and the report of the ESC. The NEPR, released in 2007, was prepared by Singapore's Ministry of Trade and Industry (MTI). The ESC report, released in February 2010, was prepared by a committee formed by the government in 2009. This committee and its eight subcommittees included representatives from the public and private sectors as well as academia.

Both the NEPR and the ESC report stress that energy security recommendations and strategies should be aligned with the larger national objectives, namely, economic competitiveness, energy security and environmental sustainability. Although there are no direct references to human security and human develop-ment, it is plausible that the government's vision of developing the country as a "global city through sustainable and inclusive growth" resonates with the core principles of human security, since human development cannot be achieved without economic growth and environmental sustainability, at least according to developmental economists and advocates of human security and development such as ul Haq and Sen.[5] This then raises the question of whether current economic security strategies can actually lead to human development and human security.

---

[4] EMA (Energy Market Authority), "Third Tank for Singapore's LNG Terminal on the Back of Strong LNG Uptake", Press Release (2 November 2010), in: http://www.slng.com.sg/UserFiles/Press/MediaRelease_ThirdTankforSporeLNG%28FINAL%29.pdf (23 December 2011).

[5] See Sen (1999) and ul Haq (1995).

An informed analysis, however, has to first appreciate the rationale behind Singapore's energy policies, both present and past, and how that is linked to the strategic role that oil has played in the nation's economic growth.

3.4.1 Petroleum as the Driver of Industrialisation

Singapore is one of the largest refining centres in the world, behind only Houston in the United States, and Rotterdam in the Netherlands. The history of oil refining in Singapore began with the installation of an oil storage facility by American oil traders in the 1870s (Horsnell 1997, p. 134). Due to its strategic location in the region, Singapore was well-placed to be a regional distribution centre. The first bulk carriage occurred on 1 July 1892, when the oil tanker *Murex* discharged 2,500 of its 4,000 tonnes and transported the rest to Bangkok. By the First World War, oil exports had reached approximately 700,000 tonnes per year. Around the same time, Singapore was, in addition to providing storage facilities, also blending and distributing oil.

After the First World War, Singapore's oil trade continued to flourish, and Singapore evolved into a port serving regional refineries such as those in Borneo and southern Sumatra (Horsnell 1997, p. 139). By the end of the 1950s, the oil trade had grown to about three million tonnes per year. By the time Singapore gained Independence in 1965, its oil industry had grown to encompass both storage and distribution, and a national refining industry had emerged.

When Singapore attained self-government in the late 1950s, the government was keen to strengthen the country's economy through industrial development and thus offered "pioneer" status to new commercial enterprises. Shell was the first company to be awarded pioneer status and began operating out of Pulau Bukom in 1961 (Horsnell 1997, p. 139). This was followed by the Japanese firms Maruzen and Toyo Menka (1962), British Petroleum (1964), Mobil (1966), Exxon (1970) and the Singapore Petroleum Company (1973).[6] Thus, by the 1970s, Singapore had transformed itself into an important regional oil refining hub, attracting the investments of major industrial players such as Shell, Exxon and Mobil.

Summing up the impact of the petroleum industry, Chang (2006, p. 124) noted that the industry was the leading sector in the country's first two decades of industrialisation, from 1965 to 1984. During this period, the Singapore economy grew an average of 7 % per annum. However, between 1985 and 2004, the electronics industry gradually replaced the oil sector as the country's leading export engine. In 1976, oil accounted for 50 % of total domestic exports. By 1984, its proportion of total domestic exports had fallen to 47.6 %. In 2001, oil accounted

---

[6] Singapore Petroleum Company Limited was incorporated in 1969 under the name Singapore Petroleum & Chemical Company (Private) Limited. See Singapore Petroleum Company, 2011, 'History and Milestones', at: http://www.spc.com.sg/aboutspc/history_milestones.asp (1 December 2011).

for just 18.3 % of total domestic exports, having been replaced by electronics which had come to represent 48 % of the total (Wu and Thia 2002, pp. 47–54). In 2010, refined petroleum products accounted for just over 21 % of the country's total exports (MTI 2011, p. iv). The value of Singapore's oil exports that same year was US$80 billion (Department of Statistics 2011).

3.4.2 Paradigm Shift in Political Thinking: Sustaining Economic Growth

A sustained increase in the price of oil, particularly in the last decade, has led to a shift in the thinking of the Singapore government. Between January 2004 and August 2006, the price of oil rose by nearly 300 %—US$75 per barrel. The global price of oil went on to reach a historic high of over US$147 per barrel in July 2008.[7] Although the price has since gone down, the current level remains higher than the first half of the last decade. As of 23 December 2011, the price of oil per barrel was US$108.22, according to the IntercontinentalExchange (ICE) Brent Crude Oil price index.[8]

> Energy economists and analysts attributed the upward pressure on the global price of oil in 2008 to several factors: the strong demand from the expanding economies of China and India; inadequate spare capacities in oil production and refining, and global concerns over "peak oil"; geopolitical uncertainties and natural disasters. The same factors have kept oil prices high in subsequent years.

Regardless of the dominant narrative among the experts, many countries were left reeling from the effects of the high oil prices. These countries suffered from slower economic growth, high inflation and high unemployment rate (these also occurred during the first and second oil shocks, in 1973 and 1979, respectively). Any deterioration in the national economy inevitably affects socioeconomic conditions. Increases in the prices of goods and services will have a serious impact on the population's quality of life, and naturally, the financially less well-off will be the most vulnerable.

Singapore, which is largely dependent on global trade and exports, was one of the countries feeling the effects of rising prices in the mid- to late-2000s. The then Deputy Prime Minister of Singapore remarked in 2006:

---

[7] Kebede, Rebekah, "Oil Hits Record above $147", in: *Reuters: US Edition* (11 July 2008), at: http://www.reuters.com/article/2008/07/11/us-markets-oil-idUST14048520080711 (23 December 2011).

[8] ICE (IntercontinentalExchange), 2011, "Brent Crude Oil", at: http://www.theice.com/ (23 December 2011).

Even as our consumption of energy has risen steadily with our rapidly growing economies, we have concurrently taken for granted that the supply of energy will continue to be cheap and sustainable. Recently, however, this idyllic state of affairs has been disrupted. (ISEAS 2006)

He announced that the government would look at energy issues from the perspectives of economic competitiveness, energy security, environmental sustainability and energy industry development. These four categories later took on a more definite shape when it was packaged as the NEPR. The government thus continued to mark out energy security as a primary driver of the national economy; its strategy had however shifted, from industrialisation to economic growth.

3.5 The NEPR: Energy for Growth

The Energy Policy Group released the NEPR themed "Energy for Growth" in November 2007. Formed in March 2006, the Energy Policy Group is an inter-ministerial group led by the MTI. The group includes representatives from the Energy Market Authority (EMA) as well as the Ministry of Finance; the Ministry of Foreign Affairs; the Ministry of the Environment and Water Resources; the Ministry of Transport; the Agency for Science, Technology and Research; the Building and Construction Authority; the Economic Development Board; the Land Transport Authority and the National Environment Agency (MTI 2007, p. 25).

The report articulates the state's holistic approach towards energy security. It outlines three main policy objectives: economic competitiveness, energy security and environmental sustainability. These three objectives translate into six strategies: to promote competitive markets; to diversify energy supplies; to improve energy efficiency; to build the energy industry, and invest in energy research and development; to step up international cooperation and to develop a whole-of-government approach.

These objectives are to be achieved by developing and strengthening government, academic and research institutions. To that end, the government introduced several agencies with specific functions aligned to the state's strategic objectives. These agencies regulate the industry and the market according to the policies set by the state. These policies and regulations in turn ensure that the energy market remains competitive and efficient. Complementing these government agencies are academic and research institutions that focus on studying energy-related issues. The impacts of the six strategies are perhaps better understood when examined against the backdrop of their implications for Singapore at the national, regional and international level, and these are discussed next.

3.5.1 Institutional Framework for Energy Security in Singapore

At the national level, the state recognises the complexity and strategic importance of a comprehensive energy policy and has thus adopted a whole-of-government approach. To that end, it has created several government agencies and a think tank, or expanded the roles of existing institutions, to fulfil specific functions (Fig. 3.1).

- *Energy Division within the MTI.* The Energy Division develops and manages Singapore's overall energy policy with the aim of supporting economic growth and addressing energy security, economic competitiveness and environmental sustainability. The division also oversees the development of the energy industry and energy research and development, and analyses energy issues using economic methodology. In addition to collaborating with other agencies, the division works closely with the EMA to set the strategic direction and policies for energy security and the liberalisation of the electricity and gas markets.
- *EMA and its Energy Planning and Development Division.* Formed in April 2001, the EMA, a statutory board under the MTI, regulates the electricity and gas industry as well as the district cooling services in designated areas.[9] Initially, the EMA's role was to liberalise the market and promote competition in the electricity and piped gas industry, and to maintain the security and reliability of the power system. Under the NEPR, the EMA's functions have been expanded to encompass strategic concerns related to Singapore's overall energy needs. It thus created the Energy Policy and Planning Division. The division has since evolved into the Energy Planning and Development Division which, among other things, plans and reviews Singapore's energy policies. The division also develops scenarios that help Singapore formulate strategic plans to secure its energy needs.
- *Clean Energy Programme Office.* Set up in April 2007, the Clean Energy Programme Office is the key inter-agency work group for planning and executing strategies to develop Singapore into a Global Clean Energy Hub. The aim is to develop clean energy products and solutions for export globally.
- *Energy Efficiency Programme Office.* Led by the National Environment Agency, the Energy Efficiency Programme Office is a multi-agency task force whose mission is to draw up a long-term plan that integrates whole-of-government efforts to improve energy efficiency, an initiative known as Energy Efficient Singapore.[10]
- *Think tanks.* The Energy Studies Institute (ESI) seeks to advance the understanding of local, regional and global energy issues through independent research and analyses aimed at addressing, informing and influencing public opinions and

---

[9] EMA (Energy Market Authority), 2010, "Overview of EMA", at: http://www.ema.gov.sg/page/40/id:94/ (1 December 2011).

[10] Energy Efficiency Programme Office, 2008, "About E2 Singapore", at: http://www.e2singapore.gov.sg/energy-efficiency-programme-office.html (1 December 2011).

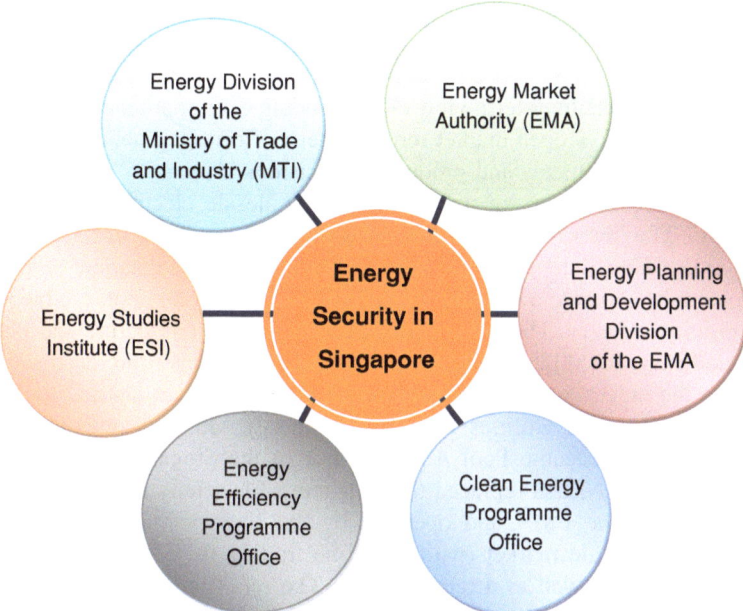

Fig. 3.1 Singapore's energy security: The national institutional framework. *Source* Based on MTI (2007)

policies.[11] The Centre for NTS Studies at the S. Rajaratnam School of International Studies analyses the impact of energy security policies on human security.[12] Other think tanks which offer valuable perspectives on energy issues include the Institute of Southeast Asian Studies, the Institute of South Asian Studies, the Lee Kuan Yew School of Public Policy and the East Asian Institute.

3.5.2 Electricity Market, Policy and Regulation

The Electricity Act of 2001 was introduced to develop a more cost-efficient electricity market through open competition and market liberalisation. When the National Electricity Market of Singapore (NEMS) opened for trading on 1 January 2003, it was Asia's first liberalised electricity market.[13]

---

[11] ESI (Energy Studies Institute), 2011, "About ESI", at: http://www.esi.nus.edu.sg (1 December 2011).

[12] Centre for NTS Studies, "Energy and Human Security Programme", at: http://www.rsis.edu.sg/nts (1 December 2011).

[13] Energy Market Company, 2008, "About the Market: Asia's First Liberalised Electricity Market", at: http://www.emcsg.com (23 August 2009).

Prior to 1995, the generation, transmission and retail segments of the electricity and gas industries were managed by the Public Utilities Board, a statutory board established in 1963. In October 1995, the electricity and piped gas industries were corporatised with the long-term view of introducing open market competition. The rationale was to allow open market forces instead of central planning to determine market prices, investments and production decisions.[14]

In 1998, the Singapore Electricity Pool, a wholesale electricity market, came into being. The market was further deregulated from 2000 onwards with the introduction of an independent system operator, the liberalisation of the retail market, and the separate ownership of the contestable and non-contestable parts of the electricity industry. The government also decided to restructure and liberalise the gas industry. In 2001, the EMA was formed. It took over the responsibility of regulating the electricity and gas industry as well as certain district cooling services from the Public Utilities Board. The Energy Market Company was also established, and it became Pool Administrator of the Singapore Electricity Pool (EMA 2009). The NEMS replaced the Singapore Electricity Pool in 2003.

Regulated by the EMA and operated by the Energy Market Company, the NEMS is a wholesale market (with real-time trading for energy, regulation and reserve products) as well as a retail market. The retail market was introduced in stages. From July 2001, consumers started to become contestable, that is, they were given the option of selecting their own retailers. Initially, however, this was only made available to consumers with a power requirement of 2 MW and above. In the next phase, consumers with an average monthly consumption of 20,000 kW h and above were given that option. The level was revised further in December 2003, to 10,000 kW h and above. These large consumers account for about 75 % of total electricity demand. At the point of writing, retail competition for the other approximately 25 %, that is, those who require less than 10,000 kW h a month, is still being studied by the EMA. Meanwhile, these small consumers continue to purchase electricity from just one source, SP Services.[15]

It appears that the intention is for the market to eventually become fully liberalised, with all industrial and domestic household consumers given the freedom to select their own electricity retailers and determine their own electricity pricing plans (EMA 2009). Households and other energy consumers will be able to exercise this freedom once the nationwide Intelligent Energy System (IES) is fully implemented (see Sect. 3.6.1 for more on the IES).

Market liberalisation could lead to market failure should power generation companies attempt to influence market prices using unfair practices. This is of particular concern in Singapore where the three largest producers, Senoko Power, PowerSeraya and Tuas Power, supply about 90 % of the nation's electricity

---

[14] EMA (Energy Market Authority), 2008, "Milestones in Restructuring (Singapore Electricity and PNG Industries)", at: http://www.ema.gov.sg (23 August 2009).

[15] EMA (Energy Market Authority), "Implementation of Electricity Vending System (EVS) Project by the Energy Market Authority", Press Release (18 October 2007).

demand (Chang 2007). To prevent the large power generation companies, or gencos, from distorting market prices by withholding supply or via other unethical practices, the EMA introduced vesting contracts. These contracts—agreements between gencos and the Market Support Services Licensee, who acts on behalf of consumers—stipulate that gencos must sell a certain quantity of electricity at a certain price (vesting price). The EMA determines the vesting price every 2 years. Vesting contracts are voluntary for smaller gencos.[16]

Certain studies have shown that the restructuring of the electricity market in Singapore does indeed have an economically and statistically significant impact on the market itself. Furthermore, vesting contracts are effective at maintaining a lower electricity price; and they also help to lower the price of electricity and its volatility during peak hours (Chang et al. 2007, p. 197).

By the end of 2008, Senoko Power, PowerSeraya and Tuas Power had been sold by Temasek Holdings, whose shareholder is Singapore's Ministry of Finance.[17] This effectively means that the Singapore government no longer owns these gencos. Senoko Power was the last of the three wholly owned gencos to be divested. These divestments, which began in 2007, were part of a larger strategy by the Singapore government to create an open and competitive power generation market. However, it remains to be seen whether, in the long run, the restructuring and liberalisation of the electricity market would lead to energy security in Singapore, and whether vesting contracts are effective at curbing the exercise of market power.

3.5.3 Research and Development

The Energy Policy Group, through the NEPR, has identified research and development as key to achieving energy security and sustainable development. The government has noted, among other projects, the use of photovoltaics as a source of clean energy, and as a means of diversifying Singapore's energy mix. The NEPR further mentions that Singapore is well-placed to develop photovoltaic capabilities for export to the region. The National Research Foundation through its Clean Energy Programme has set aside US$132 million for solar research (MTI 2007, p. 67).

---

[16] Market Support Services, 2008, "Contestability—Vesting Contracts", at: http://www.mssl.com. sg (1 August 2008).
[17] Temasek Holdings, "Temasek Sells Senoko Power to Japanese Consortium", Press Release (5 September 2008).

3.5.4 Regional Cooperation and Integration

Singapore is actively involved in various energy-related initiatives instituted by the Association of Southeast Asian Nations (ASEAN), the Asia-Pacific Economic Cooperation (APEC) and the East Asia Summit (EAS). Singapore is represented on the task forces on energy of APEC and the EAS.

As part of ASEAN Vision 2020, member states pledged to develop an interconnecting arrangement for electricity, natural gas and water through the ASEAN Power Grid and the Trans-ASEAN Gas Pipeline and Water Pipeline.[18] As a member of ASEAN, Singapore has signed the memorandum of understanding on energy security, the ASEAN Power Grid and the Trans-ASEAN Gas Pipeline.

3.5.5 International Participation

At the international level, Singapore participates actively in the UN Framework Convention on Climate Change (UNFCCC) process. Singapore was one of the 154 countries that signed the UNFCCC at the 1992 UN Conference on Environment and Development in Rio de Janeiro, Brazil. Singapore went on to ratify the UNFCCC on 29 May 1997, bringing it into force on 27 August the same year (Ministry of the Environment 2000). On 13 April 2006, Singapore acceded to the Kyoto Protocol of the UNFCCC.[19] Singapore's engagement with the UNFCCC process signifies its commitment to calibrate the country's strategies and actions with those of the global community. This would necessarily mean introducing measures aimed at reducing the country's energy intensity and consumption over the long term.

Singapore's energy policy at the regional and international level also resonates with the vision of "comprehensive energy security" articulated by former Executive Director of the IEA, Nobuo Tanaka. In a speech during Singapore International Energy Week 2011, Tanaka called for a new framework for energy security to meet the challenges brought about by the volatility of the global economy:

> The future of energy security is more complex and difficult today. It requires us to work with neighbouring countries to secure and supply energy needs in an affordable and sustainable way. I would like to see Asian countries work together in the future to create a framework that will achieve an interconnected grid in the region.[20]

---

[18] ASEAN (Association of Southeast Asian Nations), 1997, "ASEAN Vision 2020", at: http://www.aseansec.org/1814.htm (1 December 2011).

[19] Ministry of the Environment and Water Resources, "Singapore Accedes to the Kyoto Protocol", Press Release (13 April 2006).

[20] EMA (Energy Market Authority), "4th SIEW Opens with Call for Comprehensive Energy Security in Asia", Press Release (31 October 2011), at: http://www.ema.gov.sg/news/view/286 (1 November 2011).

Although Tanaka did not outline a framework for comprehensive energy security, it can be inferred from his speech that the international community needs to foster greater working relations and understanding in the areas of the cross-border flow of energy, and perhaps, of particular relevance to Singapore and Southeast Asia, in the areas of electricity imports and exports via the ASEAN Power Grid.

3.6 The ESC Report: Towards a Smart Energy Economy

The ESC was formed in May 2009 by Singapore's Prime Minister to recommend strategies for Singapore to further develop itself as a global city through sustainable and inclusive growth. Within the ESC, a subcommittee on Ensuring Energy Reliance and Sustainable Growth was formed to recommend specific strategies in support of the national objectives of economic competitiveness, energy security and environmental sustainability (that is, the same objectives as the ones outlined in the 2007 NEPR). The 2010 ESC report singles out energy security as crucial to the nation's economic competitiveness and growth, and calls for the national economy to evolve into a "smart energy economy".

The ESC report calls for the diversification of energy resources, investments in critical energy infrastructure, an increase in energy efficiency, and the use of carbon pricing schemes to protect the economy against future price hikes and in anticipation of future carbon constraints as a result of international climate change agreements. These five strategies and the associated recommendations are outlined in Table 3.2.

3.6.1 Realising the Smart Energy Economy through the IES Project

The ESC report specifies that the IES would be the central pillar of the country's smart energy economy. A pilot IES project, spearheaded by the EMA, was announced in September 2009. The IES project leverages on new communication, information and sensor technologies to enhance the efficiency and resilience of Singapore's power system, reduce wastage, shave peak loads and defer capital investments to better meet future consumer demand.[21] Table 3.3 summarises the potential benefits of the IES.

---

[21] EMA (Energy Market Authority), "Intelligent Energy System Pilot Project Kicks Off in Singapore", Press Release (29 September 2010), at: http://www.ema.gov.sg/news/view/212 (1 November 2011).

Table 3.2 Strategies and recommendations of Singapore's Economic Strategies Committee (ESC)

Strategies	Recommendations
Strategy 1: Diversifying Singapore's energy sources	1. Allow the entry of new energy options on a market basis 2. Develop renewable energy resources 3. Study the feasibility of the nuclear energy option and develop expertise in nuclear energy technologies
Strategy 2: Enhancing infrastructure and systems	4. Invest in critical energy infrastructure ahead of demand 5. Develop Jurong Island as an energy-optimised industrial cluster
Strategy 3: Increasing energy efficiency	6. Promote energy efficiency for buildings and industries, and in homes 7. Support clean and efficient technologies in transportation
Strategy 4: Strengthening the green economy	8. Establish energy as a key national research and development priority 9. Build capabilities for the green economy 10. Apply a green lens to government procurement
Strategy 5: Pricing energy right	11. Price energy to reflect its total cost

Source Based on ESC (2010)

The IES project will computerise Singapore's power grid and provide two-way communication between energy consumers and providers. Prior to the announcement of the IES project, between 2007 and 2009, the EMA had studied the feasibility of the Electricity Vending System, a retail scheme based on the use of smart-meter technology to give small-sized consumers access to a variety of electricity packages offered by different retailers. During the trial, the EMA tested how well the smart-meter technologies could be integrated with the existing national e-payment infrastructure. Smart-meter technologies form one of the key elements of the IES.

Phase I of the IES project (2010–2012) involves the development of enabling infrastructure such as smart meters and the establishment of communication networks linking energy consumers such as households and industries with energy providers. These communication networks will give energy providers and consumers access to information on energy usage. The next phase (2012–2013) will focus on the testing of the smart grid applications that will enable households equipped with smart meters to choose their own electricity pricing plans and to monitor their energy usage instantaneously through household display devices.[22] In short, then, the IES is designed to help households and industries manage their electricity consumption more efficiently.

---

[22] Ibid.

Table 3.3 Potential benefits of the intelligent energy system (IES)

Group	Potential benefits
Households	Choice of electricity retailer and pricing plan. More information to monitor and manage energy usage. Better control of major home appliances to reduce energy usage
Business (commercial and industrial consumers)	Choice of electricity retailer and pricing plan. More information for building owners and occupants to manage energy issues. Better control and automated systems at the building level to reduce energy usage
Grid owner and operator (Singapore power)	Effective communication with households and businesses to enhance delivery of electricity. Enhanced capability to detect and respond promptly to localised power outages. Easier integration of new energy sources into the grid

Source EMA (Energy Market Authority), "Intelligent Energy System Pilot Project Kicks Off in Singapore", Press Release (29 September 2010), at: http://www.ema.gov.sg/news/view/212 (1 Nov 2011)

3.7 Energy Security for Whom?

Singapore's approach to energy security, intended to drive economic growth and achieve sustainable development, appears comprehensive and balanced. However, it remains to be seen whether the liberalised market system and the smart energy economy will be sufficient to ensure that human security is achieved.

Sen (1999) and ul Haq (1995) argue that human security and development can only be achieved if they are the ends rather than the means of economic growth. They point out that wealth might not necessarily trickle down to every household, and that this could affect the ability of some households to consume basic necessities such as electricity. Therefore, they call for people to be placed at the centre of any nation's economic development and planning. Taking this point further, Sen and ul Haq note that an economic growth strategy should include a development agenda that results in the people having greater freedom of choice.

Expanding on this idea, Prahalad (2004) and Yunus (2003) suggest that people should be empowered to improve their socioeconomic conditions by participating in the market. They believe that the people's welfare could be improved through such participation, and that exclusion from the market will create and sustain socioeconomic inequalities. How well does Singapore's energy policy measure up in terms of the dimensions of choice and market participation?

According to the NEPR, the liberalisation of the national electricity market will promote an open market which would be able to price electricity competitively and thus serve the interests of both consumers and producers. However, will the benefits be felt by all if there is a wide income gap, and when the market equilibrium is determined by the energy demands of the middle or upper class?

Lower-income groups could be left behind if a fixed tariff is applied across all households. The liberalisation of the market may lead to efficient transmission of electricity, but that does not mean that all consumers would be able to afford that electricity. It should also be stressed that the notion of energy poverty for an urban population in a developed economy is different from that for a rural, agrarian society typically found in less-developed and underdeveloped economies. Those living in urban areas would have almost no recourse to alternative energy resources. Such is the case in Singapore. A majority of Singaporeans live in Housing Development Board (HDB) flats, and these residents do not as yet have access to alternatives such as biofuels or solar energy. Furthermore, unlike residents living in privately owned properties who could opt for other energy sources, HDB dwellers depend mainly on SP Services for their electricity supply. Thus, it could be argued that more attention should be paid to this group of consumers.

To address some of the challenges faced by urban low-income households in the country, the EMA and SP Services introduced the Pay-As-You-Use metering scheme in 2005. Under the Pay-As-You-Use scheme, consumers pay in advance for their electricity. This scheme is available to households who have defaulted on their payments or are in arrears. However, these households still pay the same tariff as those on the conventional scheme.

Under the smart energy economy model, poorer households could benefit from the implementation of the IES. They would have the ability to monitor their electricity usage and to choose a package that suits their needs, thus allowing them to better manage their electricity consumption. However, the smart energy economy as it is currently conceived would not necessary lead to human security for all consumers. While the IES would give consumers greater freedom of choice, it does not at present include any mechanism to make energy more affordable. To enhance energy affordability, household consumers need to be able to influence the price of electricity and to derive some profit from it.

3.8 Policy Recommendations

Prahalad's and Yunus's theoretical frameworks posit that market participation could address some of the socioeconomic challenges faced by low-income households (Sect. 3.7). Building on this stream of thought, consumers should be provided with the means to be active rather than passive participants in the energy market. Singapore's smart energy economy model could be enhanced to enable such participation.

Consumers, especially households, could be given the freedom to sell surplus electricity back to the market. Households could either sell their surplus electricity to retailers or trade surplus electricity with other households. This would have the added advantage of reducing power wastage, and addressing issues of energy surplus and overcapacity, as unused electricity would be routed back to participants needing that electricity.

For this system to be realised, new legislation that makes it possible for retailers and households to engage in the trading of electricity would have to be introduced.

The existing market-based model, which binds participants as either consumers or producers in the production and exchange of commodities, would have to be modified. The necessary technology would also have to be in place; the IES would have to be enhanced. Further, there would be a need to put in place programmes to encourage energy consumers to participate actively in the market rather than remain mere passive consumers.

This idea is not new and has been widely discussed by proponents of smart energy grids, such as Roberts (2005). If the objective of a smart energy economy is to ultimately improve connectivity, service and efficiency, then household consumers in particular should be given the opportunity to actively participate in the liberalised energy market (since the optimum price of electricity is matched, in part, against their demands). It is reasonable to argue that the smart energy economy could be designed to benefit households not just by providing them with the means to achieve energy savings, but also by giving them the ability to generate income from participating in the energy market.

3.9 Conclusion

Energy security has always been important to Singapore's economic objectives but it has in recent years taken a more prominent and central role. This could be clearly discerned from the government's 2007 NEPR and its 2010 ESC report, which identified energy as the driver of future economic development and growth. Clearly, energy security is seen as one of the building blocks of the nation's security. This appears to be a rational strategy considering how effective the petroleum industry has been in driving the country's economy forward in the last century.

However, energy security should also be understood in relation to the human security norms of human welfare and development. In particular, electricity tariffs, while determined by market forces, should be made affordable for all. Households must be able to consume electricity at a price that is not detrimental to their welfare. Essentially, this means that electricity consumption should not comprise a large portion of each household's expenditure. This is a pressing issue considering that households, particularly those in government-subsidised HDB flats, have limited access to alternative energy sources.

The IES, scheduled to be implemented in Singapore in the near future, may represent the best opportunity for the government and the people to achieve the ideals of energy and human security. A more people-centred system could be introduced in which households could be empowered with the ability to trade electricity with one another, or with industry, in the national energy market. This could help them to not only better manage their energy usage, but also achieve some measure of financial gain.

In the final analysis, energy security in the spirit of human security should position the socioeconomic well-being of human collectivities as the end rather than the means of sustainable economic development. During the global oil price hikes of 2007 and 2008, countries struggled to sustain their economies. Households in developed and developing economies alike suffered from the resulting price inflation. Standards of living dropped, and many households fell below the poverty line simply because they could not cope with the rising costs. All this happened despite the uninterrupted global supply of oil and electricity. The issue then, as it will continue to be in the future, is the ability to access energy sources at affordable prices rather than the availability of such sources.

References

Chang P-L (2006) Trade, foreign direct investment and regional competition: The case of Singapore. In: Koh WTH, Mariano RS (eds) The economic prospects of Singapore. Addison-Wesley Pearson, Singapore)

Chang Y (2007) The new electricity market of Singapore: regulatory framework, market power and competition. Energy Policy 35(1):403–412

Chang Y, Park C (2007) Electricity market structure, electricity price, and its volatility. Economics Letters 95(2):192–197

Department of Statistics (2011) Yearbook of statistics Singapore, 2011. Ministry of Trade and Industry, Singapore. http://www.singstat.gov.sg/pubn/reference/yos11/statsT-trade.pdf. Accessed 23 Dec 2011

EMA (Energy Market Authority) (2009) Introduction to the National Electricity Market of Singapore. EMA, Singapore

EMA (Energy Market Authority) (2011) Singapore Energy Statistics 2011. Research and Statistics Unit, EMA, Singapore. http://www.ema.gov.sg/media/files/facts_and_figures/fuel_mix/Fuel_Mix%20extracted%20from%20Singapore%20Energy%20Statistics.pdf. Accessed 23 Dec 2011

Emmers R, Caballero-Anthony M (2006) Introduction. In: Emmers R, Caballero-Anthony M, Acharya A (eds) Studying non-traditional security in Asia: trends and issues. Marshall Cavendish, Singapore

ESC (Economic Strategies Committee) (2010) Report of the Economic Strategies Committee. ESC, Singapore. http://app.mof.gov.sg/data/cmsresource/ESC%20Report/ESC%20Full%20Report.pdf. Accessed 19 Mar 2010

ExxonMobil (2012) 2012 The outlook for energy: a view to 2040. ExxonMobil, Irving). http://www.exxonmobil.com/Corporate/energy_outlook.aspx. Accessed 23 Dec 2011

Horsnell P (1997) Oil in Asia: markets, trading, refining, and deregulation. Oxford University Press, Oxford

IEA (International Energy Agency) (2011) Key world energy statistics 2011, http://www.iea.org/textbase/nppdf/free/2011/key_world_energy_stats.pdf. Accessed 23 Dec 2011

IEEJ (Institute of Energy Economics, Japan), Asean Centre for Energy (ACE) (2011) 3rd ASEAN energy outlook 2011. ASEAN Centre for Energy, Jakarta

ISEAS (Institute of Southeast Asian Studies) (2006) Singapore energy conference 2006: summary report. ISEAS, Singapore

Ministry of the Environment (2000) Singapore's initial national communication under the United Nations Framework Convention on Climate Change. Ministry of the Environment, Singapore

MTI (Ministry of Trade and Industry) (2006) Developing a holistic energy policy. MTI, Singapore

MTI (Ministry of Trade and Industry) (2007) Energy for growth: National Energy Policy Report. MTI, Singapore

MTI (Ministry of Trade and Industry) (2011) Economic survey of Singapore 2010. MTI, Singapore. http://app.mti.gov.sg/data/article/24221/doc/FinalReport_AES_2010.pdf. Accessed 23 Dec 2011

Ong KY (2009) Energy security in Singapore, reference paper for Ambassador Ong's address at the FES-RSIS energy security workshop, 20 October. http://www.spp.nus.edu.sg/ips/docs/pub/sp_oky_Energy%20Security%20in%20Singapore_201009.pdf. Accessed 4 Sept 2011

Prahalad CK (2004) The fortune at the bottom of the pyramid: eradicating poverty through profits. Wharton School Publishing, Upper Saddle River

Roberts P (2005) The end of oil: on the edge of a perilous new world. Mariner Books, New York

Sen A (1999) Development as freedom. Oxford University Press, Oxford

ul Haq M (1995) Reflections on human development. Oxford University Press, New York

UN (United Nations) (2010) Human security report of the Secretary-General, A/64/701

UNDP (United Nations Development Programme) (1994) Human development report 1994. UNDP, New York

UNDP (United Nations Development Programme) (2007) Overcoming vulnerability to rising oil prices: options for Asia and the Pacific. UNDP Regional Centre, Bangkok

Wu, F, Thia JP (2002) Singapore's changing growth engines since 1965: an economic history of nimble adaptability. In: Economic survey of Singapore, second quarter 2002. Ministry of Trade and Industry, Singapore, pp 45–59

Yunus M (2003) Banker to the poor: micro-lending and the battle against world poverty. Public Affairs, New York

Abbreviations

APEC	Asia-Pacific Economic Cooperation
ASEAN	Association of Southeast Asian Nations
EAS	East Asia Summit
EMA	Energy Market Authority
ESC	Economic Strategies Committee
ESI	Energy Studies Institute
HDB	Housing Development Board
ICE	Intercontinental Exchange
IEA	International Energy Agency
IEEJ	Institute of Energy Economics, Japan
IES	Intelligent Energy System
IE Singapore	International Enterprise Singapore
ISEAS	Institute of Southeast Asian Studies
LNG	Liquefied natural gas
MTI	Ministry of Trade and Industry
NEMS	National Electricity Market of Singapore
NEPR	National Energy Policy Report
NTS	Non-Traditional Security
UN	United Nations
UNDP	United Nations Development Programme
UNFCCC	United Nations Framework Convention on Climate Change

Author Biographies

Youngho Chang (Singapore) Assistant professor of Economics at the Division of Economics, Nanyang Technological University (NTU), Singapore. Since 1999, he has taught resource and energy economics, environmental economics and macroeconomics at the National University of Singapore and NTU. He has also contributed to various academic journals, including Econometric Theory, Energy Policy, Economics Letters, International Journal of Global Energy Issues and International Journal of Electronic Business Management. He received a BASc in landscape architecture from Seoul National University, South Korea; an MA in economics from Yonsei University, South Korea; and a PhD, also in economics, from the University of Hawaii at Manoa, US.

Nur Azha Putra (Singapore) Research associate at the Energy Security Division of the Energy Studies Institute, National University of Singapore. He was previously associate research fellow at the Centre for Non-Traditional Security (NTS) Studies at the S. Rajaratnam School of International Studies, Nanyang Technological University (NTU), Singapore, where he was the lead researcher for the Energy and Human Security programme and also led research on a range of other NTS issues. He holds an MSc in international political economy from NTU; and a Bachelor of Information Technology from Central Queensland University, Australia.

Chapter 4
Perspectives on India's Energy Security

Krishnan Rekha

Abstract One of the world's fastest-growing economies, India is also home to 30 % of the world's "energy poor". The result is rapidly rising energy demand that will have to be met in the face of considerable development challenges, including the need to ensure economic growth and poverty reduction while at the same time dealing with environmental impacts such as carbon emissions. The multi-dimensional nature of its energy challenge has spurred the country to go beyond earlier strategies focused on securing energy imports. It now seeks to also take advantage of the complementarities offered by renewable energy, improved energy efficiency and other sustainable energy options.

Keywords India · Energy security · Energy poverty · Energy imports · Climate change · Sustainable energy

4.1 Introduction

Though India has 17 % of the world's population, it accounts for just 5 % of global energy consumption. However, with an expanding economy and rapidly growing population, India is witnessing a soaring demand for energy. Simultaneously, a high prevalence of energy poverty and the pressure to pursue low-carbon energy alternatives could hamper the country's economic growth, already marred by large economic inequities and infrastructure-related bottlenecks.

K. Rekha (✉)
E 303, Central Park I, Sector 42, Gurgaon, Haryana, India
e-mail: rekhak.work@gmail.com

M. Caballero-Anthony et al. (eds.), *Rethinking Energy Security in Asia:*
A Non-Traditional View of Human Security, SpringerBriefs in Environment, Security,
Development and Peace 2, DOI: 10.1007/978-3-642-29703-8_4, © The Author(s) 2012

4.2 India's Energy Trends: An Overview

India's energy sector has witnessed rapid expansion in recent years, in terms of both total and per capita energy consumption. Total energy use nearly doubled between 1990 and 2008, from just 318 to 621 million tonnes of oil equivalent (mtoe); while energy use per capita increased by 45 %, from 375 to 545 kilogrammes of oil equivalent (kgoe). With India's economy growing at about 8 % annually and the country's population increasing by about 16 million each year, its energy consumption has been growing by about 4 % per year. Yet, its per capita energy consumption remains relatively low, at just a third of the world average and an eighth of the average per capita consumption in countries that make up the Organisation for Economic Co-operation and Development (OECD) (Figs. 4.1 and 4.2).

Over half of India's commercial energy comes from coal, followed by oil and gas. In the electricity sector, 54 % of the country's installed capacity is coal-based, though the share of hydropower and other renewables have inched up to 32 % (Fig. 4.3). India consumes about 160 million tonnes of crude oil, in terms of refinery throughput, and over 80 % of total petroleum consumption can be traced to imports (MoPNG 2010).

There are environmental implications along the life cycle of fossil fuels, from generation to use. Of these, climate change and its associated impacts have serious long-term and cross-border ramifications. Exploration for fossil fuels, as well as their processing, transportation and consumption, account for most of the global anthropogenic emissions of greenhouse gases (GHGs) in both developed and developing countries. India's 60 % dependence on fossil fuels compares favourably with the global average of 80 %,[1] which is partly why the country's per capita carbon dioxide emissions are so low—about 1.4 tonnes—compared to the global average of 4.6.[2] A major point of concern, however, is the pace of growth in the use of fossil fuels and the resultant growth in per capita carbon dioxide emissions.

Biomass, sourced domestically and not very widely traded (therefore referred to as non-commercial energy), is currently India's single most important source of energy, meeting a third of the country's energy needs. Nearly 625 million Indians who do not have access to modern cooking fuels continue to use firewood, chips and cow dung in traditional cooking stoves. The existing literature provides strong evidence that smoke from the combustion of solid fuels in these stoves is a significant risk factor for various diseases, predominantly acute lower respiratory infections in young children and chronic obstructive pulmonary disease in women.

---

[1] World Bank, 2011, "Energy and Mining", at: http://data.worldbank.org/topic/energy-and-mining (28 September 2011).

[2] World Bank, 2011, "CO$_2$ Emissions (Metric Tons Per Capita)", at: http://data.worldbank.org/indicator/EN.ATM.CO2E.PC (28 September 2011).

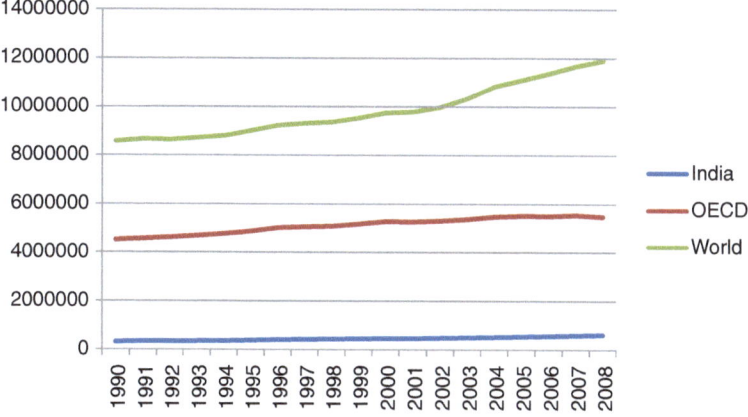

Fig. 4.1 Energy use ('000 tonnes of oil equivalent). *Source* Based on World Bank, 2011: "Energy and Mining". http://data.worldbank.org/topic/energy-and-mining Accessed 28 Sept 2011

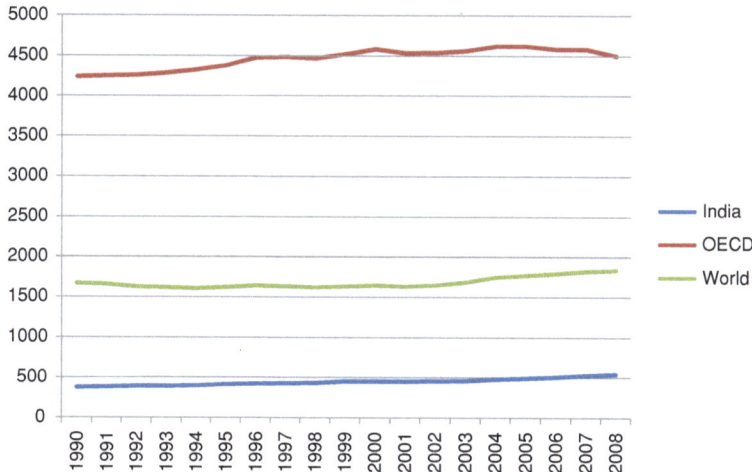

Fig. 4.2 Energy use per capita (kg of oil equivalent). *Source* Based on World Bank, 2011: "Energy and Mining". at: http://data.worldbank.org/topic/energy-and-mining (28 Sept 2011)

4.3 Understanding India's Energy Insecurities

In examining India's energy security, this chapter adopts a holistic contemporary definition of the issue: energy security comprises several aspects, including availability and affordability of energy in its various forms for all consumers, as well as environmental concerns associated with the energy sector. This is a much wider approach than the traditional supply-centric perspective, which focuses on the geopolitics of energy imports. Possible indicators of a country's energy

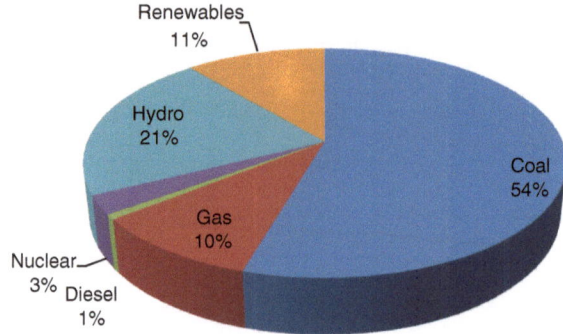

Fig. 4.3 India's installed electricity generating capacity, by fuel (July 2011). *Source* Based on Central Electricity Authority of India, "All India Region-wise Generating Installed Capacity (MW) of Power Utilities including Allocated Shares in Joint and Central Sector Utilities" (31 July 2011), at: http://www.cea.nic.in/reports/monthly/executive_rep/jul11/8.pdf (1 Aug 2011)

insecurity are therefore high dependence on imported fuels and on fossil fuels with high carbon content (Rekha 2009). Large intra-country inequities in energy access are also an indicator of energy-related insecurities. The following sections outline India's major energy security challenges.

4.3.1 Rapid Growth in Demand, and Rising Deficits

As mentioned earlier, India's energy sector is marked by rapid growth in demand. In the face of limited domestic supplies, this has resulted in large and growing energy deficits. India's average per capita electricity consumption, at 700 kilowatt-hours (kWh), is just a third of the world average. Yet, in the electricity sector, the country already suffers a shortage of nearly 8 % overall; at times of peak energy demand, the shortfall is nearly 10 % (Central Statistical Organisation 2010). The annual growth rate for electricity consumption, meanwhile, is expected to be around 7 % (CSO 2010).

With the high rate of population growth, urbanisation and economic expansion in India, and the large sections of the population still in need of greater access to electricity and modern energy sources, the country's energy demand will undoubtedly continue to rise rapidly for several decades. Projections indicate that the industrial and residential sectors will remain the largest energy users. Growth in the latter is strong, with the use of biomass decreasing significantly even while electricity consumption is rapidly increasing (De la Rue du Can et al. 2009). The fastest growth, however, is expected to be in the transport sector (De la Rue du Can et al. 2009). Given this sector's high dependence on oil, it is not surprising that projections for oil consumption indicate a 6 % annual increase till 2020, leading to the concerns discussed in the following sections.

4.3.2 Dependence on Imported Energy

About 40 % of the commercial energy consumed in India is imported. Dependence on imported oil is of particular significance given the volatility in the international oil market. Further, with the production of domestic crude oil remaining stagnant over the past 15 years, its dependence on imported oil has been on the rise. Currently, India imports about 75 % of its oil, of which nearly three-fourths is sourced from West Asia. Notably, West Asian countries (particularly Saudi Arabia, Iran and Iraq) are seeing increases in their share of India's oil imports.

India began importing natural gas in 2004, and Qatar has been the single largest supplier. (In the past 2 years, other countries such as Trinidad and Tobago and Oman have also been supplying this resource, but in negligible amounts.) Although India does have coal reserves, the country imports more than a fourth of the coal it consumes. Since 2003, imports of non-coking or thermal coal have also risen sharply, with Australia, Indonesia, China and South Africa emerging as major suppliers.

With barriers to trade having come down since the 1990s, opportunities have opened up for India to diversify its fuel mix and to begin using cleaner fuels. At the same time, trade has increased the country's vulnerability to international energy-related price shocks. To a certain extent, the rise in Indian oil prices since 2002 shows how price increases in the international market have been able to move into the domestic market. In addition, the burden due to subsidies and other forms of under-recoveries (the difference between cost and realised price) has remained large. India's significant dependence on oil, particularly imported oil, makes it highly susceptible to the adverse effects of rising oil prices, despite part of this vulnerability being buffered by the country's economic resilience (United Nations Development Programme 2007 based on The Energy and Resources Institute 2007).

4.3.3 Energy Poverty

Despite large strides having been made in the Indian energy sector, the issue of access to energy remains worrisome. Rao et al. (2009) provide a few relevant indicators:

- Over 40 % of India's population lack electricity access and, hence, use kerosene for lighting.
- India's average per capita consumption of electricity is estimated at around 700 kWh, less than a third of the estimated world average.
- Only one-sixth of those using electricity consume over 100 kWh per month (compared to the US per capita average of 900 kWh per month).
- About 74 % of rural India and 21 % of urban India continue to rely on biomass-based fuels for cooking.

Even more disheartening is the fact that there has been so little change in this situation, despite a flurry of infrastructure, technology and policy initiatives. Further, an analysis of survey data from 2002 to 2008 (reported by the National Sample Survey Office in 2010) reveals a widening rural–urban divide. The share of rural households using "dirty" cooking fuels has remained at 85 %, while the figure for urban households has declined from 28 to 23 %. In urban areas, the number of households using liquefied petroleum gas (LPG) as a primary fuel source increased by more than 10 % (from 50 to 61 %), while rural households using LPG increased only marginally, from 8.1 to 9.1 %. In urban areas, this was accompanied by a marked decline in the use of kerosene for cooking (from 15.3 to 7.6 %), while in rural areas this figure went down from 2 to 0.6 %. Though subsidies for both LPG and kerosene were intended to help alleviate energy poverty in India, this has not taken place, largely due to leakages and mis-targeting of subsidies (discussed in Sect. 4.5). Individual choice over fuel usage is also influenced by factors such as fuel availability, culture, geography, the extent of urbanisation and the design of state policies.

India has in recent years invested heavily in projects aimed at improving access to electricity. In April 2005, a mammoth rural electrification programme, the Rajiv Gandhi Grameen Vidyutikaran Yojana (RGGVY, the Rajiv Gandhi Scheme for Rural Electricity, Infrastructure and Household Electrification), was launched. The programme merged the various schemes aimed at ensuring the electrification of all villages and other areas of habitation. The RGGVY also used a new definition of village electrification that included the development of basic electricity infra-structure, the provision of electricity to public places (such as schools and health centres) as well as a minimum electrification of at least 10 % of total households in any given village. Prior to this, a village was considered electrified if power was used within its revenue area, regardless of the number of households electrified, the amount of electricity consumed or the purpose of its consumption.

Despite this improved definition, significant gaps remain, as indicated in an analysis of consultative reviews of the programme (Krishnaswamy 2010; Green-peace 2011). There are problems in terms of the number of hours that electricity is supplied every day, the amount of power made available, the end uses that can be supported, and the extent to which access to electricity is facilitating a transition away from inefficient and unhealthy conventional fuels (e.g. traditional biomass for cooking or kerosene for lighting). Importantly, there has been a failure to mainstream decentralised renewable energy sources into the RGGVY, despite the programme having a component (albeit small) for decentralised distributed generation.

4.4 India's Efforts to Improve Energy Security

If we examine India's energy security from a broad perspective, we will find that a number of problems continue to persist. This is due not to an absence of efforts to address these issues, but to the country's large energy deficits rendering such steps inadequate. In some instances, the action taken has been poorly conceptualised or executed, as in the case of the RGGVY. A sometimes confusing spectrum of organisations, dominated by ministries and government agencies (see Box 4.1), are engaged in various aspects of the energy sector.

Several recent policies and pieces of legislation have strengthened the regulatory framework for various aspects of India's energy sector. In late 2008, India's cabinet approved the Integrated Energy Policy (IEP). Introduced by the Planning Commission of the Government of India, the IEP is a landmark document in at least two ways: first, it seeks to integrate the diverse segments of India's energy sector and, second, it brings together the relevant environmental, developmental and economic goals that are the pillars of energy security. The IEP seeks to develop "the ability to supply lifeline energy to all citizens irrespective of their ability to pay for it as well as meet their effective demand for safe and convenient energy to satisfy their various needs at competitive prices, at all times and with a prescribed confidence level considering shocks and disruptions that can be reasonably expected" (TERI 2008). As such, the IEP's aim is to address all aspects of energy security—energy use and supply, availability, energy imports, efficiency, affordability, pricing as well as environmental concerns. Broadly, India's energy security efforts can be classified as follows:

- measures to reduce risks associated with imported energy, including the development of alternative energy sources to reduce dependence on fossil fuels (which have high carbon content and will increasingly have to be imported).
- measures to improve energy efficiency, which can help to reduce both environmental damage and demand for scarce energy resources.
- measures to improve people's access to energy, particularly electricity and clean cooking fuels.

Box 4.1: Ensuring Energy Security in India: A Complex of Institutions and Agencies

The Indian energy sector comprises a plethora of organisations, both governmental and non-governmental, each playing roles in policymaking, regulation and operations. Various departments in the government deal with different energy sources. Energy security additionally requires engagement with other departments such as the Ministry of External Affairs, the Ministry of Finance, the Ministry of Rural Development and the Ministry of Industry.

Policymaking Institutions
In India, energy-related policymaking is steered by the Planning Commission along with the Ministry of Power, the Ministry of Coal, the Ministry of Petroleum and Natural Gas, the Ministry of New and Renewable Energy, the Department of Atomic Energy and agencies affiliated with these bodies.

To enable a systematic approach to policy formulation, the Energy Coordination Committee was constituted in July 2005.[3] It promotes coordination between departments and provides institutional support for decision-making in the areas of energy planning and security. The prime minister chairs the committee, and members include representatives of relevant agencies and ministries.

Regulatory Agencies

Electricity falls under the concurrent list of the Constitution, meaning that it is administered by both central and state governments.

The Central Electricity Regulatory Commission, constituted under the Electricity Commissions Act (1998), serves as the regulator at the central level. It aims to promote competition, efficiency and economy in the bulk power markets, improve the quality of power supply and promote investment. It also protects consumer interests, particularly by advising the government on the removal of institutional barriers in order to bridge the gap between demand and supply.

The State Electricity Regulatory Commission serves as the regulatory body for individual states.

The Petroleum and Natural Gas Regulatory Board was established in June 2007. It is tasked with regulating the refining, processing, storage, transportation, distribution, marketing and sale of crude oil, petroleum products and natural gas. It aims to protect the interests of consumers as well as entities engaged in these areas, and to ensure uninterrupted, adequate supply of these products throughout the country. Note, however, that the production of crude oil and natural gas is outside the Board's purview.[4]

Energy Sector Ownership and Operations

The coal sector is dominated by state-owned enterprises. Coal India Limited accounts for over 84 % of domestic coal produced. Private sector companies are involved in just 7 % of coal production, which is exclusively for captive (i.e. earmarked) uses.

In the oil sector, there are several large state-controlled companies. The private sector, however, plays a large role in this sector.

The majority of the power sector (nearly 75 %) is also state-owned, through the state electricity boards and a few companies. Power Grid Corporation of India, a state-controlled company that has interconnections with Nepal and Bhutan, is responsible for high-voltage bulk inter-state transmission of power.

---

[3] Prime Minister's Office, n.d., "The Energy Coordination Committee", at: http://pmindia.nic.in/eccbody.htm (5 June 2008).

[4] Petroleum and Natural Gas Regulatory Board, n.d., "Objective of the Board", at: http://www.pngrb.gov.in/ (5 June 2008).

4.4.1 Managing Dependence on Energy Imports

India is a major and growing trader in the global energy market, and it has been affected by the international anxiety over volatile oil prices and possible disruptions in the oil supply. India is among the world's largest importers of oil and coal. Over the last few years, dependence on imported coal has increased due to the increasing gap between demand and supply, as well as due to the low and deteriorating quality of domestic coal. India is not a significant trader in natural gas; and though imports of natural gas are expected to grow, India's energy imports are likely to continue to be dominated by oil followed by coal. The country will increasingly need to tap the global market to meet its coal and oil needs.

For the most part, India's attempts to mitigate price and supply risks associated with energy imports have taken the form of investments in other countries, negotiations for cross-border pipelines, and long-term contracts in oil and gas. On the investment front, Indian oil companies have established a presence in several countries in Asia, Africa and Latin America, and the country also holds a stake in Russia's substantial Sakhalin-1 block (oil and gas fields). However, the extent to which such investments help to secure energy is debatable both in terms of the effectiveness of the strategy and its relevance for various energy forms.

To secure gas, India is in the midst of negotiations for three major transnational pipeline projects: Iran-Pakistan-India (IPI), Turkmenistan-Afghanistan-Pakistan-India (TAPI) and Myanmar-Bangladesh-India (MBI). Talks on the IPI and TAPI have been slow due to geopolitical tensions in the areas involved, and the prospects for the MBI are complicated by the fact that Myanmar allocates much of its gas to China. Therefore, India is likely to continue to import liquefied natural gas (LNG). India already has LNG terminals at Dahej, Hazira, Dabhol and Kochi, and possible sources for long-term contracts other than Qatar are Australia, Indonesia, Malaysia, Algeria and Nigeria. In the LNG market, India has to compete with Japan, Korea and China, and price negotiations will be critical.

Because the last few years have been marked by spikes in oil prices, there has been renewed interest in two options seen as critical to mitigating risks associated with volatility in the international oil market. Policymakers have been keen, first, to build strategic petroleum reserves and, second, to identify alternative fuels for the transport sector (MoPNG 2010), which in India accounts for nearly 35 % of petroleum consumption. Storage tanks are under construction for India's 5 million tonne strategic petroleum reserves, and the country has plans to increase its reserves to about 132 million barrels by 2020.[5]

---

[5] Sharma, Rakesh, "India Unveils Strategic Oil Stockpile Plans", in: *The Wall Street Journal* (21 December 2011), at: http://online.wsj.com/article/SB10001424052970204464404577111893998 225190.html?mod=googlenews_wsj (10 January 2012).

India's biofuel strategy stipulates that over the period 2012–2017 the country will produce vegetable oil sufficient to achieve a 20 % blend with diesel (Planning Commission 2003). A recent development is the creation of a National Policy on Biofuels, which suggests fiscal incentives for biofuels, calls for a certification mechanism to ensure compliance with required blending rates, and emphasises the need for the indigenous production of bio-diesel feedstock from non-edible oil seeds grown on wastelands, and degraded and marginal lands. While liquid biofuels do hold significant promise, the response to this option has been mixed. It is being highlighted that biofuels development needs to proceed with caution, bearing in mind issues such as trade-offs between food and fuel.[6]

There is some interest in India in a process in which coal is converted into liquids such as gasoline or diesel, referred to as coal-to-liquids (CTL). An official decision has been taken to open up the country's coal reserves for use by CTL plants, and the Ministry of Coal has begun inviting companies interested in setting up such plants in India to apply to use the country's captive coal mines (those allocated for a specific need), which amount to up to 1.5 billion tonnes of reserves. Yet while this policy came into force in 2007, CTL is still viewed with some caution due to its heavy use of resources (particularly water and energy), its carbon-related implications, as well as India's rising dependence on imports of high-quality coal (TERI 2010).

4.4.2 Improving Energy Efficiency

In India, industry and transport are the largest consumers of energy, accounting for 44 % and 17 % of the energy consumed respectively. However, the residential and commercial sectors, which together account for 15 % of energy consumption, are growing rapidly. Each of these sectors offers significant opportunity for improvements in energy efficiency. Although India's agriculture sector is not particularly energy intensive (accounting for around 7 % of energy consumed), it still has the potential for energy savings through improvements in irrigation technologies and practices.

India's commitment to energy conservation goes back to the 1970s, when the Petroleum Conservation Research Association was created with a mandate to reduce dependence on oil. The Energy Conservation Act (2001) addresses some market failures by providing for minimum energy standards to be set, energy-consumption labels to be placed on appliances and equipment, and conservation-minded building codes to be promulgated. It also allowed officials to push for energy-use monitoring, verification and reporting by large-scale energy users, and the establishment of energy-consumption norms for these consumers.

The Bureau of Energy Efficiency was set up under the Energy Conservation Act, with the mission of reducing India's energy intensity (the energy efficiency of

---

[6] Several of these issues were raised at the Dialogue on Biofuels and Energy Security, a seminar organised by The Energy and Resources Institute (TERI) in August (2007).

its economy). Yet energy efficiency has progressed unequally in India. While large-scale industrial enterprises have made rapid strides in response to market signals and due to industry restructuring (which entailed consolidation and the entry of efficient global players), the more informal and dispersed small, medium and micro enterprises remain an area of significant concern.

Globally, barriers to energy efficiency include inadequate access to capital, insulation from price signals, information asymmetry (when one party is unwilling to share information) and split incentives (when parties entering into a contract have different goals) (OECD 2007). In India, major barriers appear to stem from similar factors (Sathaye 2006):

- pricing and subsidies that are not reflective of efficiency or environmental implications thus reducing the effectiveness of price signals in influencing energy demand.
- lack of delivery mechanisms for energy-efficiency services.
- policy and knowledge barriers on specific interventions.
- failure to treat energy efficiency on the same economic basis as new capacity.

These are also some of the issues that form the cornerstone of the National Mission on Enhanced Energy Efficiency, approved by the Prime Minister's Council on Climate Change in 2009. A number of initiatives are underway, as set out by the Mission (see Box 4.2).

Box 4.2: Initiatives under the National Mission on Enhanced Energy Efficiency

Perform, Achieve and Trade (PAT)
Conceptualised by the National Mission on Enhanced Energy Efficiency, PAT is an important initiative. It focuses on energy-intensive industries, setting goals for energy efficiency and facilitating these industries in achieving these goals, then granting tradable energy permits to those who achieve the targets. Industries that fail to do so are required to buy these permits or pay penalties.

Market Transformation for Energy Efficiency (MTEE)
Under the MTEE initiative, accelerated shifts to energy-efficient appliances in designated sectors will be enabled through a variety of measures, including financing through the UN's Clean Development Mechanism (CDM)[7] wherever possible.

---

[7] The Clean Development Mechanism (CDM) is a flexible mechanism by which certain ("Annex 1") countries under the Kyoto Protocol of the Intergovernmental Panel on Climate Change (IPCC) can meet a part of their carbon caps using Certified Emission Reductions from carbon emission reduction projects in developing countries.

Financing
The Mission will foster the creation of mechanisms to help finance demand-side management through a revolving fund (to promote carbon finance) and a partial risk-guarantee fund (to provide commercial banks with partial coverage of risk exposure against loans made for energy-efficiency projects).

Technology
A technology strategy for the power sector is also being drawn up, focusing on a variety of new technologies: fluidised bed combustion (FBC), super-critical and ultra-supercritical technologies as well as integrated gasification combined cycle (IGCC) demonstration plants.[8]

Institutional Strengthening
Parallel processes of institutional strengthening have also been implemented to kick-start and manage these initiatives. These include the creation of a public sector company to facilitate progress on the above-mentioned issues. A stronger Bureau of Energy Efficiency will also continue to play a quasi-regulatory role, strengthening the agencies identified to take up energy-efficiency initiatives at the state level, and promoting awareness at multiple levels.

4.4.3 Harnessing Alternative Energy

4.4.3.1 Renewable Energy

Over the last five years, the trebling of India's capacity for renewable energy has been due largely to policy and regulatory changes, including those provided for in the Electricity Act (2003). The major driving forces include the fixed targets that have been set for procurement of renewable energy and the preferential tariffs for meeting those targets. Two other external forces—volatile oil prices and rising concerns over climate change—could act as incentives for the further development of "green energy" options.

---

[8] In fluidised bed combustion (FBC), coal is burned in a reactor comprising a bed through which gas is fed to keep the fuel in a turbulent state. This improves combustion, heat transfer and recovery of waste products. FBC systems improve the environmental impact of coal-based electricity, reducing certain emissions by 90 %. Supercritical and ultra-supercritical technologies operate at increasingly higher temperatures and pressures and therefore achieve higher efficiencies than conventional units. Integrated gasification combined cycle (IGCC) plants use a gasifier to convert coal (or other carbon-based materials) to syngas, which is then cleaned and used in a gas turbine to produce electricity. Waste heat from the gas turbine is recovered to create steam, which drives a steam turbine, producing more electricity.

As part of its policy to promote renewable energy, India launched the Renewable Energy Certificate (REC) Mechanism in 2010. Under this scheme, generators of renewable energy are granted an REC per unit of green energy that they contribute to the grid. India also has in place a scheme called the Renewable Purchase Obligation (RPO). The RPO sets the minimum percentage of electricity consumption that must come from renewable energy sources. Each state has its own RPO level, decided upon by the State Electricity Regulatory Commission. RECs have begun to be traded on the power exchanges, with the RECs being purchased by states that are otherwise unable to meet their RPO.

However, several factors hinder the development of technologies for renewable energy and these would have to be addressed (TERI 2010):

- *Inter-state variations in the potential for renewable energy* Given the difficulties in the physical transfer of electricity, the REC Mechanism would aid states in meeting their RPO.
- *Attitudinal issues* There is apprehension relating, for instance, to the intermittent nature of wind energy. However, such fears appear exaggerated given the negligible (2 %) contribution that wind currently makes to electricity generated in India. Further, wind is able to contribute as much as 20 % of the electricity generated in other countries, such as Denmark.
- *Multiplicity of agencies involved* The central and state regulatory commissions, the Ministry of New and Renewable Energy (and organisations affiliated to it), the Department of Science and Technology, the Ministry of Power, the Indian Renewable Energy Development Agency and state nodal agencies all need to come together. There is also a need to enhance availability of information on policy responses and on the institutional mechanisms providing access to resources from the Indian Renewable Energy Development Agency, multilateral agencies and foreign direct investment.

India has ambitious targets for renewable energy. These include increasing the capacity to generate renewable energy by 40–55 gigawatts (GW), by the end of the 13th Five-Year Plan (2022); and a 1 % annual increase in capacity for the generation of renewable energy, as set out in the National Action Plan on Climate Change. Particularly ambitious is the Jawaharlal Nehru National Solar Mission, which sets a target of adding 1 GW of solar capacity between 2010 and 2013, and increasing combined solar capacity (plants using solar thermal collectors) from 9 megawatts (MW) in 2010 to 20 GW by 2022. The World Bank (2010) points out that to add 40 GW of renewable energy by 2022, India will have to meet the ambitious targets set out by the Jawaharlal Nehru National Solar Mission, double its wind-power capacity, quadruple its small-hydropower capacity, fully realise co-generation (or combined heat and power) capacity, and increase the use of biomass by a factor of 5–6. It is imperative, therefore, that the barriers mentioned above are tackled with a sense of urgency.

4.4.3.2 Nuclear Energy

India's nuclear capacity is currently around 4,000 MW, but this could increase to 20,000 MW by 2020. Though the country's three-stage nuclear programme is eventually aimed at reducing dependence on nuclear-related imports (see Box 4.3), in the near term the plan calls for securing supplies of uranium, largely through imports. While uranium exploration within India has been expanded, this remains at an exploratory stage, and in the meantime the country will have to utilise the international market. Following the signing of the US–India Civil Nuclear Cooperation Agreement in 2008 and a subsequent waiver given by the international Nuclear Suppliers Group, India entered into import pacts for nuclear fuel with France, Russia, Kazakhstan and Canada, and is looking to sign similar agreements with Tajikistan, Namibia, Mongolia, Brazil, South Africa, Gabon and Australia. Fears over reaching "peak uranium"—when remaining reserves could begin to decline—are likely to raise competition for the resource throughout the world.

There are other obstacles to India's nuclear programme. First, the Civil Liability for Nuclear Damage Bill (2010), or the Civil Nuclear Liability Bill, was passed in the Parliament but has yet to be notified. The Bill, which defines the scope of civil liability for nuclear damage, is an important step in activating the US–India deal and will be critical in funnelling investment into the sector. The extent of the liability, however, has been a matter of much debate.

Second, even once a liability regime is created, stringent security measures to prevent nuclear proliferation will still be critical. This is an even more significant issue in the aftermath of the Fukushima nuclear disaster in Japan in March 2011, as well as the public protests over proposed nuclear plants at Jaitapur and Kudankulamin India. The Nuclear Power Corporation of India has recently undertaken the re-evaluation of 20 power plants, those in operation as well as those under construction. But whatever the eventual results of the reviews, it will not be easy to secure public acceptance of these projects. With the possible bifurcation of India's nuclear programme into military and civilian branches, it now appears probable that an independent nuclear regulatory authority will be created.[9]

Box 4.3: India's Nuclear Programme in Brief

India's nuclear energy strategy, involving a three-stage programme, is visionary in its attempt to reduce dependence on imported fuel.

The Three Stages
In the first stage, the programme calls for using uranium to produce power and plutonium. During the second stage, the latter is to be used as a "driver" fuel along with thorium (of which India has one of the world's largest

---

[9] Mohan, M.P. Ram, "How Safe Is India's Nuclear Energy Programme", in: *livemint.com* (23 August 2011), at: http://www.livemint.com/2011/08/22202845/How-safe-is-India8217s-nucl.html (accessed in 1 November 2011).

reserves) to produce uranium-233. During the third stage, power is to be produced using this artificially produced Uranium-233.

Progress
The first stage of this plan is already well underway. The country's first 540 MW indigenous reactors are already online, and the construction of larger, 700 MW, indigenous reactors is currently taking place. The second stage is now in development, with India's first commercial-scale fast-breeder reactor expected to be in operation soon. The third stage is in the design phase, and the Indian government is planning to start construction of a 300 MW advanced heavy-water reactor (AHWR), which will make use of domestically available thorium. India will move on to the third stage once its fast-breeder reactors reach significant capacity.

4.5 Priorities Going Forward

In recent years, the Indian energy sector's regulatory and institutional framework has undergone several positive changes. Nevertheless, many challenges lie ahead, and these will have to be dealt with through bold and direct approaches.

In particular, the lines between energy policy and domestic politics must be clearly drawn. In order to meet India's burgeoning energy demand while managing the implications for the environment, transformations are required in all three facets of the energy sector: domestic supply, import and demand. In creating new incentives for domestic change in terms of both demand and supply, energy policy, particularly with regard to prices, can provide critical signals for both efficiency and fuel choice.

Typically, however, energy prices have been steered by political rather than economic considerations, and successive coalition governments have only made matters worse. The costs of irrational energy pricing and of mis-targeted subsidies are too high to be ignored (for examples of price-induced distortions in the oil sector, see Box 4.4).

Apart from pricing, the orientation of several schemes towards specific targets has also led to their failure. Most electrification programmes have shown much greater focus on the number of villages electrified, for instance, rather than on ensuring that the benefits from electrification are maximised.

Box 4.4: Price-induced Distortions in India's Energy Sector

Imposes high cost burden
With domestic prices for diesel, kerosene and LPG kept below import-parity prices despite the partial deregulation of the oil sector, the bill incurred for oil subsidies is huge. A large fraction of that bill is borne by oil companies in

the form of "under-recovery". For instance, currently under-recovery on kerosene supplied through the Public Distribution System (known as PDS kerosene) is Rs 11.30 (US$0.22) per litre, while the government's subsidy is Rs 0.82 (US$0.02). It is reported that oil marketing companies are currently incurring daily under-recovery of Rs 38.8 million (US$753,658) on the sale of diesel, PDS kerosene and LPG.[10]

Benefits the rich
Seen as an alternative fuel for households dependent on traditional fuel, LPG is highly subsidised. Yet estimates indicate that 40 % of the LPG subsidy is enjoyed by just 6.75 % of the population, and these consumers are from the country's highest income groups (TERI 2007).

Irrelevant for the poor
The poor, for whose sake subsidies are ostensibly continued, are often unable to gain access to subsidised kerosene and LPG because they lack ration cards and/or official LPG connections in the absence of valid proof of residence. A recent study of poor households reinforces previous findings that, in India, the poor have to rely on the thriving unofficial (black) market where kerosene and LPG are sold at three to four times the official rate. The study also shows that while the official price of LPG went up by 30 %, the unofficial price increased by more than 50 %. Figures for kerosene are even more worrisome, with a 1 % rise in the official price having taken place alongside a 30–60 % rise in the unofficial price (TERI 2007).

Leads to product misuse
Estimates show that about 25–40 % of kerosene is siphoned off and used for adulterating diesel, with drastic economic and health consequences.

Encourages inefficient fuel use
Although highly subsidised, kerosene is used mostly for lighting, a very inefficient use for this fuel.

Discourages introduction of clean energy options
Solar photovoltaic systems, for instance, is not attractive as a power back-up option given that the price of kerosene is currently heavily subsidised. However, such options could become financially attractive if kerosene subsidies were to be removed (as suggested in TERI 2007).

---

[10] Petroleum Planning and Analysis Cell, "Petroleum prices and under-recoveries" (9 January 2011), in: http://ppac.org.in/WRITEREADDATA/PS_oil_prices.pdf (9 January 2012).

"Energy diplomacy" has increasingly been seen in India's various negotiations, such as those regarding gas pipelines and investment in energy infrastructure overseas. With the country's external relations increasingly involving an energy dimension, an energy security unit was established in the Ministry of External Affairs in September 2007 and upgraded to a full-fledged division in 2009. This signals the mainstreaming of energy in India's diplomacy efforts, and the move is expected to bring about more constructive and significant engagements between India and the global energy system.

The Energy Security Division now supports India's international engagement through sustained diplomatic interventions. This includes the country's relations with international energy agencies such as the International Renewable Energy Agency, the International Energy Agency (IEA) and the International Partnership for Energy Efficiency Cooperation (MEA 2010). The division also supports the efforts of corporate entities, both in the public and private sectors, in acquiring energy assets overseas, in transferring new and emerging technologies to India, and in building strategic partnerships with foreign companies.

The 2008 release of the country's National Action Plan on Climate Change remains a significant step, through which enhanced focus on energy efficiency and renewable energy will have positive spin-offs for energy security. The National Mission on Enhanced Energy Efficiency (Sect. 4.4.2) and the Jawaharlal Nehru National Solar Mission (Sect. 4.4.3.1) are two of the eight bodies envisaged within the National Action Plan on Climate Change, and the establishment of several others also have implications for the development of the Indian energy sector.

Significant obstacles remain in India's transition towards higher energy security, and recent developments point to the need for an increasingly responsive domestic energy system. Though the sector is now largely deregulated, parts of it remain confined to government-owned enterprises, and a shake-up in the policy and institutional context will almost certainly be necessary. Ministries and agencies that currently operate in silos will need to open up considerably. Decentralised governance—a larger role for the states and also for local authorities—can play a catalytic role. The decentralisation of governance will lead to a host of benefits, including a deeper understanding of local needs, greater community ownership, better speed in decision-making and increased flexibility. Faced with the significant problem of energy poverty, India's rapidly expanding energy sector is simultaneously constrained by the problem of carbon emissions and the need for more energy resources. Reconciling these varied and sometimes diverse goals will remain a challenge.

References

Central Statistical Organisation (CSO) (2010) Economic Survey 2009–10. Government of India, New Delhi

De la Rue du Can S, McNeil M, Sathaye J (2009) India energy outlook: end use demand in India to 2020. Ernest Orlando Lawrence Berkeley National Laboratory, California. http://ies.lbl.gov/espubs/india_energy_outlook.pdf Accessed 28 Sept 2011

Greenpeace (2011) RGGVY social survey report: Madhubani district, Bihar. http://www.green
    peace.org/india/Global/india/report/RGGVY-_Madhubani_Report%20FINAL.pdf   Accessed
    28 Sept 2011
Rekha K (2009) How energy secure is South Asia: observations through an environmental lens.
    Energy Environ 20(6):967–972
Krishnaswamy S (2010) Shifting of goal posts—rural electrification in India: a progress report.
    Vasudha Foundation, New Delhi
Ministry of External Affairs (MEA) (2010) Annual Report 2009–10. Government of India,
    New Delhi
Ministry of Petroleum and Natural Gas (MoPNG) (2010) Basic statistics of Indian petroleum and
    natural gas 2009–10. Government of India, New Delhi
National Sample Survey Office (2010) Household consumer expenditure in India, 2007–08.
    Government of India, New Delhi
Organisation for Economic Co-operation and Development (OECD), International Energy
    Agency (IEA) (2007) Mind the gap: quantifying principal–agent Problems in Energy
    Efficiency. http://www.iea.org/textbase/nppdf/free/2007/mind_the_gap.pdf Accessed 28 Sept
    2011
Planning Commission (2003) Report of the committee on the development of bio-fuel.
    Government of India, New Delhi
Rao N, Sant G, Rajan SC (2009) An overview of Indian energy trends: low carbon growth and
    development challenges. Prayas, Energy Group, Pune. http://www.climateworks.org/
    download/?id=f21a4576-0cec-4ee3-bd3f-86d2acd578ce Accessed 28 Sept 2011
Sathaye J (2006) Implementing end-use efficiency improvements in India: drawing from
    experience in the US and other countries, Paper for the US–India energy efficiency
    conference, New Delhi, India, 2 May 2006
The Energy and Resources Institute (TERI) (2007) Policy study on impact of rising oil prices on
    the poor and implications for achievement of the MDGs. http://www.teriin.org/upfiles/
    projects/ES/2005RD24forrmated_20080920152942.pdf Accessed 28 Sept 2011
The Energy and Resources Institute (TERI) (2008) Mitigation options for India: the role of the
    international community. http://www.teriin.org/events/docs/Cop14/mitigation.pdf Accessed
    28 Sept 2011
The Energy and Resources Institute (TERI) (2010) Building an Energy Secure Future for India: in
    consultation with stakeholders. http://www.teriin.org/ResUpdate/Building_an_Energy_Secure
    _Future_for_India_(2010).pdf Accessed 28 Sept 2011
United Nations Development Programme (UNDP) (2007) Overcoming vulnerability to rising oil
    prices: options for Asia and the Pacific. UNDP, Bangkok
World Bank (2010) Unleashing the potential of renewable energy in India. http://www-
    wds.worldbank.org/external/default/main?pagePK=64193027&piPK=64187937&theSitePK=
    523679&menuPK=64187510&searchMenuPK=64187511&entityID=000356161_201102230
    02615&cid=3001 Accessed 28 Sept 2011

Abbreviations

AHWR	Advanced heavy-water reactor
CDM	Clean development mechanism
CTL	Coal-to-liquids
FBC	Fluidised bed combustion
GHG	Greenhouse gas
GW	Gigawatt
IEA	International energy agency
IEP	Integrated energy policy
IGCC	Integrated gasification combined cycle

IPCC	Intergovernmental panel on climate change
IPI	Iran-Pakistan-India (pipeline)
kgoe	Kilogrammes of oil equivalent
kWh	Kilowatt-hour
LNG	Liquefied natural gas
LPG	Liquefied petroleum gas
MBI	Myanmar-Bangladesh-India (pipeline)
MTEE	Market transformation for energy efficiency
mtoe	Million tonnes of oil equivalent
MW	Megawatt
OECD	Organisation for economic co-operation and development
PAT	Perform, achieve and trade
REC	Renewable energy certificate
RGGVY	Rajiv Gandhi scheme for rural electricity, infrastructure and household electrification (Rajiv Gandhi Grameen Vidyutikaran Yojana)
RPO	Renewable purchase obligation
TAPI	Turkmenistan-Afghanistan-Pakistan-India (pipeline)

Author Biography

Rekha Krishnan (India): She researches independently on energy and resource sustainability. An economist with over 15 years of experience in the energy sector, her areas of work include energy security, energy policy, energy poverty, resource economics and development studies. Rekha has played a lead role in various major projects of the United Nations Development Programme (UNDP), Asian Development Bank (ADB) and United Nations Environment Programme (UNEP). She has been a discussant in many international conferences and has several publications to her credit. She is currently working on a book on India's energy challenges and is pursuing a number of initiatives on resource efficiency and equality.

Chapter 5
Beyond Efficiency: China's Energy Saving and Emission Reduction Initiatives vis-à-vis Human Development

Yuxin Zheng and Sofiah Jamil

Abstract China's recent energy efficiency and emissions reduction efforts are overviewed in this chapter along with their implications and achievements. While China's efforts have been highly ambitious and productive, there have also been several challenges, as increased energy efficiency is only part of the solution for ensuring sustainable development. The chapter puts forth that China could achieve its energy saving targets more reasonably and effectively if alternative approaches were considered, heeding the need to deepen and broaden the concept of energy intensity, amending existing regulations on energy use, extending energy saving measures to areas of consumption and exploring alternative development approaches to traditional ways of industrialisation.

Keywords China · Consumption · Development roadmap/strategy · Emission reduction · Energy efficiency · Regulation of energy use

Y. Zheng
Institute of Quantitative and Technical Economics,
Chinese Academy of Social Sciences, No. 5, Jianguomennei Street,
Beijing 100732, China
e-mail: zhengyuxin@cass.org.cn
URL: www.iqte1.cass.cn/english/home.htm

S. Jamil (✉)
Centre for Non-Traditional Security (NTS) Studies, S. Rajaratnam School
of International Studies (RSIS), Nanyang Technological University (NTU),
Block S4, Level B4, Nanyang Avenue, Singapore 639798, Singapore
e-mail: issofiah@ntu.edu.sg
URL: www.rsis.edu.sg/nts

M. Caballero-Anthony et al. (eds.), *Rethinking Energy Security in Asia:*
A Non-Traditional View of Human Security, SpringerBriefs in Environment, Security,
Development and Peace 2, DOI: 10.1007/978-3-642-29703-8_5, © The Author(s) 2012

5.1 Introduction

In the lead up to the United Nations Framework Convention on Climate Change (UNFCCC) meeting in Copenhagen in 2009, China pledged to reduce its carbon intensity by 40–45 % by 2020 in response to increasing international pressure to reduce its carbon emissions. Central to seeing this pledge to fruition is China's 12th five-year plan, which was launched in 2011 and aims to reduce the country's carbon intensity by 17 %.[1] The 12th five-year plan marks the official start of China's low carbon economy and provides an important indicator of China's social and economic development. Measures in the plan will reduce fossil fuel consumption by promoting the development and application of green technology and renewable energy. While such plans may reflect a more responsible attitude from China with regard to climate change mitigation, its efforts to develop a low carbon economy are largely a result of domestic demands for a change in its mode of economic growth and a strategic decision to reduce dependence on imported oil.

Given its ambitious plans, the development of a low carbon economy in China will undoubtedly face several challenges—primarily due to its rapid pace of industrialisation and urbanisation, which generates a high level of carbon intensity, that is coupled with the limited domestic sources of energy to support such growth. As such, the transition to a low carbon economy would require not only a breakthrough in new energy developments but also energy saving and emission reduction policies that can effectively be implemented.

This chapter will provide a Chinese perspective on China's efforts to increase energy efficiency and reduce carbon emissions. Following a brief review of China's recent energy saving and emissions reduction efforts, the chapter will analyse some of the challenges to these efforts that have arisen in recent years. Finally, it concludes with a brief overview of findings and some suggestions on alternative means of meeting its energy consumption targets more reasonably and effectively.

5.2 Rapid Economic Growth: Limitations and Fallouts

China is the largest developing country in the world and is home to a fifth of the world's population. These demographics, together with the reform and liberalisation policies set in motion in the late 1970s, have paved the way for China's rapid economic growth, which is reflected in key indicators, such as its gross domestic product (GDP) that in 2010 (China Statistics Press 2011, p. 47) was 20.6 times that in 1978. Such rapid GDP growth was instrumental in China jumping from 15th

---

[1] "The Outline of 12th Five-year Plan of Economic and Social Development of People's Republic of China", at: http://www.ce.cn/macro/more/201103/16/t20110316_22304698.shtml (20 February 2012).

position, in terms of countries with the highest GDPs in the world, to second place, not too far behind the US. Given China's rapid economic growth and the accelerating rate of industrialisation and urbanisation, its energy demands have increased steadily, particularly with the rapid expansion of its energy-intensive and highly polluting industrial sectors. Such brisk growth, while resulting in an unprecedented scale of economic activity, has also brought with it three key detriments: rising resource scarcity, extreme pollution and increasing carbon emissions.

5.2.1 Resource Scarcity

China's insatiable appetite for energy is clear from its energy consumption in recent years, which has made it one of the world's largest energy consumers alongside the US. Certain statistical figures have even suggested that China had in 2009 surpassed the US in terms of energy consumption. For instance, according to the International Energy Agency (IEA), China's energy use in 2009 was equivalent to 2.265 billion tonnes of oil, which was more than the 2.17 billion tonnes that the US used in the same year, thus making China the world's largest energy consumer in 2009 (IEA 2010). In any case, China's energy consumption could potentially increase over time and overtake the US, as China's energy use per capita is currently only one-fifth that of the US while its current GDP per capita is less than one-tenth of the US.[2]

Despite the high level of energy consumption in China, sections of Chinese society continue to experience energy poverty. In rural regions of China, for instance, a considerable number of farmers have no sustainable access to energy sources and still rely on biomass, such as straw and firewood, as a primary source of energy. This, in some cases, is also contributing to deforestation where farmers are forced to rely on forest resources for energy due to a lack of biomass. There is also the added pressure of rising energy consumption levels in urban areas due to China's rapid rate of urbanisation, which increases by about 1 % annually and accounts for nearly ten million farmers relocating to cities every year. Interestingly, the average energy use per person in townships is about 3.5 times that in the countryside. While China has one of the best rates of electrification in the world (Pan et al. 2006), its increasing urban population will certainly also cause a substantial increase in energy demand.

What makes the challenge of meeting such rising energy demands more acute is the limited availability of domestic energy reserves in China—particularly for oil and gas. China's petroleum reserves are expected to meet its demands for only the next 10.7 years, reflecting a high disparity from the ratio of storage volume to exploitation per year for global petroleum, which is currently at 45.7 years

---

[2] In 2010, GDP per capita for the US was US$47,284 while that for China was US$4,382; see IEA, "IEA Member Countries", at: http://www.iea.org/country/index.asp (29 February 2012).

(BP 2010, p. 6). Rapid economic growth has increased China's dependence on oil imports to the point that, since 1993, China is a net importer of oil. In 2010, for instance, China surpassed the US in terms of the rate of oil imports—China imported 55.2 % of its crude oil needs in 2010 while corresponding figures for the US were 53.5 %.[3] Where China was importing about one-third of its oil needs in 2000, its oil imports are expected to rise to nearly 80 % by 2030 (IEA 2007, p. 329). Where coal is concerned, in spite of having a relatively large coal reserve, China's ratio of storage volume to exploitation per year is only 38 years, which is also just a third of the average global ratio of storage volume to exploitation per year.[4] Coal, nevertheless, does provide China the opportunity to reduce its dependence on the world energy market in light of its rising energy consumption. A transformation of China's economic development strategy is thus unsurprisingly a priority for both energy and state security.

5.2.2 The Pollution Problem

The second challenge that China faces as a result of its rapid economic growth is severe environmental pollution. China's rapid industrialisation process has achieved in some dozen years what developing countries usually take centuries to accomplish, and this has led to an extensive array of environmental challenges. China's intense use of coal energy, which accounts for 70 % of its total energy use, is the root cause of much of its problems—coal is highly polluting and releases concentrated elements of carbon following direct combustion.

Pollution in China has significant implications for human health. In recent years, the extent of China's air pollution has been comparable to that seen during the most serious polluting period in developed countries—the 1950s and 1960s (Rawski 2009, p. 27).[5] In addition, China has one of the worst cases of water pollution in the world. Pollution accidents occur frequently causing extensive damage to human health and the environment and, due to this, treatment of traditional pollutants is imperative for China. As such, China would need to prioritise the reduction of traditional pollutants within its emission reduction initiatives, so as to not only mitigate climate change but also, more importantly, ensure the sustainability of its environmental and societal well-being.

---

[3] Chang, Xiaohua; Li, Xinmin, "China Surpassed the US in Terms of the Rate of Oil Imports", in: *Xinhua News Agency* (13 August 2011), at: http://news.ifeng.com/mainland/detail_2011_08/13/8386819_0.shtml (26 January 2012).

[4] The average global ratio of storage volume to exploitation per year is 119; see "China's Ratio of Coal Storage Volume to Exploitation per Year is Far Lower than Average Level in the World", in: *Chinahexie.org* (22 February 2011), at: http://www.chinahexie.org.cn/a/zixun/redianshiping/20110221/7451.html (20 February 2012).

[5] Rawski, Thomas G., "Urban Air Quality in China: Historical and Comparative Perspectives", at: http://www.econ.pitt.edu/papers/Thomas_air_quality.pdf (20 February 2012).

Although China's environmental condition has generally worsened, there have been some signs of improvement since the 1990s as a result of environmental monitoring mechanisms that were initiated in the 1970s. However, these improvements have been limited to special areas or specific domains, such as big cities or highly developed regions. Moreover, air pollution in China's cities continues to fluctuate.

5.2.3 Substantial Greenhouse Gas Emissions

Closely associated with the issue of pollution is the rising level of carbon emissions in China. For instance, the PBL Netherlands Environmental Assessment Agency (PBL) reported in June 2007 that China became the world's leading carbon emitter in 2006, with its carbon emissions surpassing those of the US for the year.[6] Similarly, the IEA estimated China's carbon dioxide (CO_2) emissions in 2009 to be higher than that of the US.[7] Also, although China's per capita energy use and per capita carbon emissions still remain much below those of the US,[8] the rapid rise in China's energy use and CO_2 emissions make it a key contributor of global greenhouse gas (GHG) emissions.

While the exact extent of climate change remains uncertain and requires more in-depth study, particularly at the regional and subregional levels, the international community agrees that the threat of climate change is real and needs to be addressed. As an emerging major world power and a significant contributor of carbon emissions, China will have to bear the responsibility of reducing its emissions in the near future. It must seriously consider strategic policies and technical solutions that reduce carbon emissions in order to avoid paying further costs in the future.

---

[6] PBL Netherlands Environmental Assessment Agency, "China now No. 1 in CO_2 Emissions; USA in Second Position", at: http://www.pbl.nl/en/dossiers/Climatechange/moreinfo/Chinano wno1inCO2emissionsUSAinsecondposition (20 February 2012).

[7] CO_2 emissions in 2009 for China and the US were 6831.60 metric tonnes (Mt) and 5195.02 Mt, respectively; see IEA, "Selected 2009 Indicators for China, People's Republic of", at: http://www.iea.org/stats/indicators.asp?COUNTRY_CODE=CN (29 February 2012); IEA, "Selected 2009 Indicators for United States", at: http://www.iea.org/stats/indicators.asp?COUNTRY_CODE=US (29 February 2012).

[8] Energy use/capita figures in 2009 for China and the US were 1.70 tons of oil equivalent (toe) and 7.03 toe, respectively. Similarly, CO_2/capita figures in 2009 for China and the US were 5.13 tons (t) and 16.90 t, respectively; see IEA, "Selected 2009 Indicators for China, People's Republic of", at: http://www.iea.org/stats/indicators.asp?COUNTRY_CODE=CN (29 February 2012); IEA, "Selected 2009 Indicators for United States", at: http://www.iea.org/stats/indicators.asp?COUNTRY_CODE=US (29 February 2012).

5.3 Energy Conservation and Emission Reduction: Government Initiatives

The Chinese government has made efforts to address the above-mentioned concerns by attempting energy conservation. A systematic energy conservation work plan was initiated in the 1980s in China, according to which development would be ensured even as conservation efforts remained a priority. The chapter will next examine some of the key initiatives and policy changes that were effected in China to support energy conservation and emission reduction as well as their implications for the country.

5.3.1 Measures: Structural, Technological and Political

China's accelerated industrialisation and urbanisation have stimulated substantial investments in fixed assets and resulted in the exponential growth of its heavy and chemical industries, all of which have caused an increase in its energy intensity levels, particularly during the 2002–2004 period. In response to this increase in energy intensity, China strengthened regulations over energy use during its 11th five-year plan, which established for the first time quotas to significantly reduce energy consumption intensity (set targets would lower energy consumption by 20 % in 5 years). The 11th five-year plan also made energy conservation and emission reduction a focus of macroeconomic regulation and an important area for the transformation of China's current mode of economic development. A series of strong policies and measures for promoting energy saving and emission reduction were introduced.

As a means of bringing in structural changes for energy conservation, the Chinese government introduced a number of policies to guide changes in the industrial sector. This included the use of financial, taxation and price policies to promote the development of energy-efficient industries, curb high energy consumption industries and restrict the export of products that required high energy consumption. Changes were also made to the standards established for energy consumption in backward enterprises just as a policy of 'increase the large and decrease the small'[9] was implemented to promote technological progress and accelerate the elimination of inefficient production capabilities.

Where technological innovation for energy efficiency was concerned, the government initiated key energy efficiency projects in as many as 1,000 enterprises that played an exemplary role in promoting energy conservation strategies at all

---

[9] The policy of "increase the large and decrease the small" (上大压小 [Shang da ya xiao] in Chinese) refers to state support for the construction of large-scale production equipment in highly energy-efficient and forceful enterprises that would wash out small-scale production equipment low in energy efficiency.

levels of local government. A substantial decrease was witnessed in the energy consumption per unit of main product as a result of such policies. The implementation of key projects for energy efficiency led to several positive developments as: (i) it provided experience in energy efficiency to the industrial sector; (ii) it established standards of energy consumption per unit for products in key enterprises, which would then serve as the standards for products entering the global markets; (iii) pilot projects on recycling were implemented in some iron and steel enterprises; and (iv) increased investments in energy efficiency by the central government resulted in similarly high investments in energy efficiency schemes at the local government and enterprise levels.

The above initiatives were supported by changes at the political organisational level too. For instance, China revised its Law of Energy Conservation in 2007 and established monitoring mechanisms for energy efficiency, including the National Energy Conservation Center (NECC). These would help to facilitate the following: (i) clearly stating mandatory regulations on government procurement of energy saving products; (ii) encouraging the professionalisation of energy conservation services for small- and medium-sized enterprises (SMEs); (iii) carrying out a wide range of energy efficiency awareness campaigns; and (iv) strengthening the capabilities in implementing energy efficiency initiatives.

5.3.2 Outcome

The initiatives described above have allowed China to make significant inroads in terms of energy conservation. For instance, although China's economic growth quadrupled by the end of the last century, its energy consumption only doubled. During the 1980–2006 period, China's national economic growth at 9.8 % was supported by energy use that only grew at an average rate of 5.6 % annually. At 2005 prices, energy use per 10,000 Yuan of GDP dropped from 3.39 tons of standard coal (tce) in 1980 to 1.21 tce in 2006, thus realising an annual energy saving rate of 3.9 %.[10] Increased efficiency in energy processes, conversion, storage and terminal usage also resulted in energy consumption of up to 33 % in 2006 compared to 25 % in 1980.[11] Likewise, energy consumption per unit product in China decreased significantly even as the integrated energy consumption of steel, cement, large-scale synthetic ammonia and other products greatly improved, thereby further narrowing the gap between China and the developed world.

During the period of the 11th five-year plan specifically, energy consumption per unit of GDP decreased by 19.1 %, in effect meeting the targets stipulated in the plan.

---

[10] Information Office of the State Council of the People's Republic of China, *China's Energy Conditions and Policies*, December 2007, at: http://en.ndrc.gov.cn/policyrelease/P0200712275 02260511798.pdf (26 January 2012).

[11] Ibid.

Table 5.1 China's energy use per unit of GDP during the 11th five-year plan (GDP at 2005 prices)

Year	Energy use per unit of GDP (tce/10,000 Yuan)	Rate of energy saving (%)
2005	1.276	–
2006	1.241	2.74
2007	1.179	5.04
2008	1.118	5.20
2009	1.077	3.61
2010	1.034	4.01

Source China Statistics Press (2011: pp. 47, 259)
GDP gross domestic product, *tce* tons of standard coal equivalent

Table 5.2 China's primary energy use and composition during the 11th five-year plan

Year	Total energy use (10,000 tce)	Weight of total energy use (%)		
		Coal	Petroleum	Natural gas
2005	235,997	70.8	19.8	2.6
2006	258,676	71.1	19.3	2.9
2007	280,508	71.1	18.8	3.3
2008	291,448	70.3	18.3	3.7
2009	306,647	70.4	17.9	3.9
2010	325,000	70.9	16.5	4.3

Source China Statistics Press (2011, p. 259)
tce tons of standard coal equivalent

From 2006 to 2010, for instance, China steadily reduced its annual energy consumption per unit of GDP. Significant reductions were seen in 2007 and 2008, reflecting energy consumption per capita of at least 5 % (Table 5.1). China's annual primary energy usage continued to increase during the 11th five-year plan alongside its economic growth (Table 5.2), with energy consumption in 2010 reaching 3.25 billion tce (equivalent to the total energy consumed in 2002 and 2003). However, the dependence of economic development on energy use also showed a simultaneous decrease, as seen from the declining annual elasticity of energy consumption (Table 5.3), suggesting that China's energy saving policies were largely successful.

The statistics above indicate that China's total energy mix continues to be highly dependent on coal, with the proportion of coal used for total primary energy consumption remaining at around the 70 % mark. Given this trend, it will be difficult for China to substitute coal as the main source of energy in the short term. Nevertheless, China's efforts to diversify into other energy resources are begetting positive results, as shown from its decreasing oil consumption. In China, oil was increasingly substituted with natural gas and renewable energy during the 11th five-year plan. China's efforts toward improving its energy efficiency and elec-trification have also borne positive results, as seen from the higher annual elasticity

Table 5.3 China's economic growth and elasticity coefficients of energy consumption during the 11th five-year plan

Year	Growth of energy use compared to previous year (%)	Growth of electric power use compared to previous year (%)	Growth of GDP compared to previous year (%)	Elasticity of energy use	Elasticity of electric power use
2005	10.6	13.5	11.3	0.93	1.19
2006	9.6	14.6	12.7	0.76	1.15
2007	8.4	14.4	14.2	0.59	1.01
2008	3.9	5.6	9.6	0.41	0.58
2009	5.2	7.2	9.2	0.57	0.79
2010	5.9	13.1	10.3	0.57	1.27

Source China Statistics Press (2011, p. 263)
GDP gross domestic product

coefficients of electricity use when compared with the elasticity coefficients of energy use.

Environmental pollution and CO_2 emissions have decreased as a result of China's significant efforts toward energy efficiency and emission reduction. As 30 % of total energy consumption in China is accounted for by energy consumption in buildings, the 11th five-year plan initiated a refurbishment drive of nearly 180 million square meters of residential buildings, endowing them with energy efficiency features, that required an investment of 24.4 billion RMB Yuan.[12] It is expected that about 40 million tce will be saved and a reduction of emissions amounting to nearly 100 million tons of CO_2 and over eight million tons of sulphur dioxide effected during the service life of these reconstructed buildings.[13] The positive outcomes seen for these energy efficiency efforts have prompted similar drives for other old residential areas as well. Currently, approximately 1.2 billion square metres of old housing have been marked up for reconstruction by 2020 in China, at an estimated cost of 300 billion RMB Yuan.

China's policy of 'increase the large and decrease the small', whereby small-scale plants and factories that were not productive were replaced by large-scale plants, has helped to reduce lack of productivity and so reaped positive results. Small-scale plants that were replaced by large-scale advanced-technology equipment—with a consequent reduction in the energy consumption of products—included small thermal power (76.82 million kW; nearly equivalent to Britain's installed capacity), iron (120 million tons), steel (72 million tons) and cement

---

[12] Ministry of Finance People's Republic of China, "Further Promoting the Energy-saving Reconstruction of Northern Building", in: *China State Finance* (2011), No. 16, at: http://czzz. mof.gov.cn/mofhome/czzz/zhongguocaizhengzazhishe_daohanglanmu/zhongguocaizhengzaz hishe_kanwudaodu/zhongguocaizhengzazhishe_zhongguocaizheng/33455/3345/789/201110/t20 111026_602136.html (20 February 2012).

[13] Ibid.

(370 million tons) plants.[14] Installed capacities also increased as a result of this drive. For instance, small thermal power plants (72 million kW) that were shut down between 2006 and 2010 were replaced by larger units with much higher capacities (270.93 million kW). The standard coal consumption per kilowatts hour (kW h) has also gradually reduced from 370 g in 2005 to 333 g in 2010—a 10.0 % reduction in 5 years.[15] This is significant, as China's energy consumption in terms of electricity has, in many respects, achieved a high degree of international standards.[16]

Ongoing efforts will potentially decrease China's emissions further. Similar to the 11th five-year plan, the 12th five-year plan also targets to significantly reduce energy consumption per unit of GDP (by 17 % in 5 years), in keeping with China's commitment in 2005 to reduce its carbon emissions per unit of GDP by 40–45 % by 2020 in a staggered manner. In August 2011, the State Council officially issued work plans on energy saving and emission reduction for the 12th five-year plan as well as a breakdown of national and provincial targets.[17] There is little doubt that the Chinese government is committed to energy saving and emission reduction and, given this, it is likely that China will achieve the targets it has set for itself through the 12th five-year plan and keep to its commitment of reducing carbon emissions per capita by 2020.

5.4 Challenges

The Chinese government is determined to see through its ambitious and tough energy efficiency and emission reduction targets and plans. However, there are several concerns that have not yet been fully considered and therefore pose challenges to China's efforts in this direction. The chapter touches upon on these issues next.

---

[14] "National Development and Reform Commission (NDRC) Announced that China Achieved the 11th Five-year Plan Target of Washing Out Backward Production Capacity", in: *Xinhua News Agency* (1 October 2011), at: http://www.chinanews.com/ny/2011/10-01/3367063.shtml (20 February 2012).

[15] Policy Study Department of National Development and Reform Commission, "Respective of Energy Saving and Emission Reduction in 11th Five-year Plan" (27 September 2011), at: http://www.sdpc.gov.cn/xwfb/t20110927_435642.htm (20 February 2012).

[16] China Electricity Council; US Association of Environment Protection, "Research on Emission Reduction for China Electricity Sector", at: http://huanzi.cec.org.cn/dongtai/2011-11-22/74594.html (20 February 2012).

[17] The Central People's Government of the People's Republic of China, "The Comprehensive Energy Saving and Emission Reduction Program for 12th Five-Year Plan ("十二五" 节能减排综合性工作方案)", in: *Bulletin of the State Council* (2011) 26, at: http://www.gov.cn/zwgk/2011-09/07/content_1941731.htm (20 February 2012).

5.4.1 Sustaining the Short-term Drop in Energy Consumption Intensity

A sudden reduction in energy intensity in the short term can have a series of adverse effects. Economic trends have demonstrated that long-term energy consumption intensity of most developed countries is reflected in a concave curve—the initial increase is followed by a decrease when industrialised progress is achieved. It should be noted that while the long-term trend demonstrates monotonous increase and decrease over time, short-term trends do not abide by this rule and are therefore ambiguous and uncertain. In other words, while energy consumption intensity does significantly decline during the middle phase of industrialisation, fluctuations of energy consumption intensity may also occur as a result of uncertainties of economic activity, which is a normal phenomenon too.

The energy intensity trends in China reflect the above reasoning—the general trend in energy intensity has been on the decline since the late 1970s except for the 2002–2004 period, which saw a spike in energy intensity. A series of developments may have contributed to the latter, including: (i) development strategies for China's western regions and the rapid development of the real estate industry as a result of reforms in housing policies; and (ii) China's entry into the World Trade Organization (WTO), which resulted in an increase in foreign trade and accelerated its urbanisation, the scale of infrastructure construction and the development of heavy and chemical industries. In light of these factors, it may be difficult to simply categorise the temporary rise in China's energy consumption intensity during the 2002–2004 period as an abnormality.

Endogenous factors influence energy consumption intensity based on the stage of economic development and the economic situation at a particular time. It may therefore be inappropriate to set short-term targets for levels of macro energy use and expect economic activities to abide strictly within these ranges with little room or leeway for adjustments. In fact, from a macro point of view, short-term energy efficiency quotas should be avoided as much as possible and short-term fluctuations given due consideration when setting long-term targets for energy efficiency and emission reduction.

The targets set by China for lowering energy consumption intensity during the 11th five-year plan were essentially achieved. Should its energy consumption intensity continue to drop substantially during the 12th five-year plan as well, China would have realised a continuous steady decline in energy consumption intensity for ten straight years. While the possibility of this happening cannot be discounted outright, it may be prudent to prepare for fluctuations in energy consumption intensity. Evidence suggests that energy consumption intensity during the 12th five-year plan period has varied widely.[18] Government responses to such

---

[18] Miao Wei, the Minister for Industry and Information in China, said on 11 November 2011 that "The target of reducing energy consumption per industrial value added this year is 4.5 %, we should finish 3.4 % in first 3 quarters, however just 2.56 % finished in fact, because the large

situations warrant careful consideration, as China will likely have to pay a heavier cost should it seek to merely adopt more severe administrative measures to iron out these fluctuations.

5.4.2 Reducing Existing Energy Consumption and Carbon Intensity Levels

There are inherent limitations to achieving targets on energy saving and emission reduction, as ultimately the law of economics cannot be denied. For instance, China's efforts to restrict energy consumption during its 11th five-year plan even went so far as some local governments halting electricity supply in their drive to rein in electricity use. In spite of such measures affecting normal production and everyday life, China's energy consumption intensity could only be reduced by 19.1 %, thus missing its target of 20 % reduction. The likelihood of this trend being amplified is high, as China is still in a stage of preliminary capital accumulation and has the potential to grow even more if it continues on the conventional industrialisation path as followed by developed countries.

There are also limits to how much reduction in carbon emissions can be gained from energy saving initiatives. Where the climate change discourse is concerned, energy efficiency is usually associated with the fossil fuel industry. Conversely, it follows that were all sources of energy either sustainable or renewable, energy saving initiatives would be limited to merely accounting for reduced production costs—a scenario that could be addressed via market mechanisms alone. In this case, the higher the proportion of consumption of renewable energy, the more relaxed should the government's efforts be in controlling energy consumption (including the energy consumption intensity index).

China could achieve a low level of energy consumption intensity in one of two ways. Firstly, were China's energy consumption intensity gradually reduced in the short term, capital accumulation would occur slightly faster (China's high savings ratio provides the preconditions), given the higher effectiveness of investments. China's energy consumption intensity would as a result be greatly reduced and it would reach the levels of developed countries more quickly. Secondly, a relative but substantial drop in energy consumption intensity in the short term would affect the rate of capital accumulation. This in turn would prevent the energy consumption intensity from dropping further by augmenting the time needed to finally reach the energy intensity levels of developed countries. China's industrialisation process would in this scenario be lengthened.

(Footnote 18 continued)

demand on energy intensive products"; see "The First Three Quarters of Industrial Energy Saving and Emission Reduction "Loans"", in: *Xinhua News Agency* (12 November 2011), at: http://news.xinhuanet.com/2011-11/12/c_122270187.htm (26 January 2012).

While the provisions of the Kyoto Protocol do not require developing countries to undertake obligations on reductions in GHG emissions, it would be in China's interests to seize the opportunity to actively create an environment that fosters low energy consumption. Then again, it would not be wise for China to simply pursue a strategy that promotes substantial drop of energy consumption intensity in the short term, as energy saving technologies and realising low energy or low carbon consumption require early inputs, including in the form of energy resources. No doubt, such a strategy lacks in foresight and could adversely further the reduction of energy consumption intensity for China in the future. Indeed, current developments in energy efficiency or low carbon technologies are likely to not operate effectively in the absence of subsidies for economic activities that are carbon intensive.

The world energy system is in the preliminary stages of shifting to non-fossil fuel based energy sources even as the development of new energy technologies has only produced minimal effect. However, several reports are optimistic about the increasing relevance of alternative energy sources in the decades to come. For instance, the development of renewable energy was cited as a 'most ambitious idea' in a report by the United Nations (UN), which also predicted that these would account for three-quarters of total energy consumption by 2050.[19] The WWF, on the other hand, was even more optimistic in noting that the human population could be nearly totally dependent on renewable sources for electricity, transportation, industries and household use by 2050 (WWF et al. 2011). The Intergovernmental Panel on Climate Change (IPCC), meanwhile, suggested that renewable energy would meet 17–78 % of total energy demand in China.[20] Despite the optimism on the future role of renewable energy sources, the existing energy saving practices of states and societies (and more so in developing countries) are not as active as claimed. Such optimistic prospects therefore need to be taken with a pinch of salt, and it is necessary that all options for developing energy sources for the future are thought through and any problems and uncertainty associated with such development be mitigated effectively.

---

[19] Ning, Baoying (Transl) (Ed), "2050 Renewable Energy will Become the World's Leading Energy", in: *Research Dynamic Monitoring: Climate Change Science Album* (2011) 10: 9. For the original article, see AFP, "Renewables Major Part of 2050 World Energy Mix: UN", in: *The Independent* (10 May 2011), at: http://www.independent.co.uk/environment/renewables-major-part-of-2050-world-energy-mix-un-2281695.html?origin=internalSearch (29 February 2012).

[20] China's non-fossil fuel based energy is expected to account for 11.4 % of total energy used by 2015, 15 % by 2020 and 40 % by 2050; see Du, Xiangwan, "Developing New Energy and March on the Road of New Type of Industrialization", in: *Chinese Social Sciences News* (2 August 2011): 5.

5.4.3 Consumerism

Despite China's considerable efforts toward energy saving and emission reduction, consumer behaviour and their perceptions on development have not changed. What is more, due to a popular perception that energy consumption intensity is equivalent to energy efficiency, reduced energy consumption intensity is often simply equated with energy savings. Such misconceptions need to be addressed and alternative modes of thinking cultivated.

In China, a high energy consuming lifestyle is very much desired, to the point that the consumption of luxury items has even exceeded that of developed countries in many aspects. The tendency to give in to a consumerist lifestyle in China is further exacerbated by the fact that China has become in recent years more integrated into the world economic system. The massive amounts of international capital entering China allows its production and consumption patterns to rapidly integrate with those of more developed countries. What is more, the desire to catch up with the level of development in developed countries is still a benchmark for many in the country.

The main challenge for China is to find ways of weakening the level of foreign influence on the consumption behaviours of the Chinese middle- and high-income groups. Consumer demand is an important aspect of a country's macroeconomic policies as well as a significant component of China's long-term strategies. However, there is some inconsistency between consumer demand stimulation and resource saving—a contributing factor is that, with the rise of capitalism and consumerism, societies have for the longest time grossly underpriced goods and services, thereby externalising the costs of production and ensuring that more could be consumed for less. China must address these imbalances between production and consumption in order to ensure sustainable development.

It is, therefore, absolutely necessary that China incorporates measures that strengthen its ability to manage consumer demand, including a set of guidelines that controls consumption and shifts some of the externalised costs to the consumer. This could be made applicable for various forms of consumption, be it the consumption of energy sources or material goods. When applying to the consumption of energy resources, however, conscientious care should be taken to prevent wider inequalities with regard to the more vulnerable segments of society. A progressive form of pricing, where wealthier segments of the Chinese society are allocated a premium price but those less privileged pay a lower, more affordable rate, might be worth considering.[21]

Yet, initiatives such as progressive pricing may be disadvantageous in some respects or have limitations. For instance, wealthier sections of the Chinese society that are able to afford premium prices may persist with consuming high levels of energy resources in spite of prices being prohibitive. Although consumers are at

---

[21] "Progressive Pricing", in: *China Daily* (12 October 2010), at: http://www2.chinadaily. com.cn/bizchina/2010-10/12/content_11399475.htm (26 January 2012).

liberty to increase their consumption, it would not be desirable to allow consumption to increase to the extent of unlimited indulgence that also increases energy use in China. While China continues to encourage consumer demand, guidelines on how to avoid falling into a trap of excessive consumerism should be put in place. China's economic development programme lacks such a policy on consumption and does not have clarity on how to overcome the shortcomings of the market economy, including the paradox of thrift. There is also a dearth of in-depth research on how to encourage sparing or frugal behaviour.

It is imperative that alternative solutions that effectively reduce the consumption of energy and general materials be sought. Equally, tackling the issue of waste should be a priority even for privileged government circles as a means of enhancing transparency and limiting corruption. A combination of these two measures may perhaps contribute to China's energy saving and emission reduction initiatives more effectively.

5.5 Industrialisation Versus Socioeconomic Development: A Rethink of the Traditional Roadmap

In view of the above-mentioned challenges to the effective reduction of energy consumption and carbon emissions, there is a need for a broader analysis of China's overall economic development, as most of its energy saving and emission reduction initiatives have concentrated on the production sectors, mainly aiming to eliminate backward production capacity, improve production technology and energy efficiency, and curb the development of high energy consumption industries. Indeed, although China's energy saving and emission reduction initiatives are part and parcel of its efforts to transform the conventional mode of its development, much of it still follows the traditional path that has hitherto been adopted by developed countries.

Logically speaking, developed countries that have low energy intensities would make appropriate models for China to learn from. However, the traditional path of development, as seen in these countries, is wealth oriented, and such a development path has not only increased resource scarcity and environmental degradation but also perpetuated a decay of social conscience. Many developed countries have, in truth, had to reconsider this tortuous path since the 1960s and are devoting more attention to issues of social development.

Closer home, in Asia, the earthquake in Japan on 11 March 2011 and the subsequent tsunami and nuclear crisis, wherein flaws in nuclear safety monitoring mechanisms could have been overcome with greater foresight, served as a profound warning of the consequences of prioritising economic growth and development over holistic considerations for human life. Equally, it calls attention to the possibility of huge risks and threats being downplayed as highly unlikely in an endless pursuit of material gains. Although highly technological projects may seem to be innovative solutions to existing problems, it is important to remain wary of the motives of interested groups behind such projects that may, by and large, be profit oriented with little regard for societal concerns. Such profiteering disregards the non-traditional security (NTS) approach, where individuals and communities are of central concern. It is thus necessary to weigh both the tangible and intangible costs of such projects by taking into account the views of various stakeholders across multiple levels.

There is also a need to reflect on the paradoxical trend where, despite an increase in economic growth, there is growing socioeconomic inequality in developing societies. While it is unrealistic to cease economic development, developing countries, such as China, should review their development policies and examine the extent to which economic growth has aided in the socioeconomic development of its people. Limited resources should be effectively utilised for safeguarding the common man and ensuring his basic needs, be it education, healthcare and sanitation needs, sufficient food and clean water, or access to electricity and other modern forms of energy. Ensuring these basic needs—as highlighted in the UN's Millennium Development Goals—would help communities and individuals to achieve human security in its various aspects (UNDP 1994).[22] Reducing the gaps in equality would also help to keep in check the exploitation of resources by the rich for their own interests and promote its utilisation for public interests instead. That societies with high inequalities are likely to pay little attention to resource conservation should also be of concern to China, which has a large population.

An NTS approach that highlights human development would provide a useful framework for broadening the analysis of China's overall economic development. Human development, according to the United Nations Development Programme (UNDP), entails expanding people's choices, building on shared natural resources and requires that sustainability at multiple levels be addressed through methods that are equitable and empowering (UNDP 2011). Further substantiating the importance of this approach, the latest Human Development Report highlights the challenges to be faced in meeting sustainable and equitable progress and underscores that "environmental degradation intensifies inequality through adverse impacts on already disadvantaged people and how inequalities in human development amplify environmental degradation" (UNDP 2011). In keeping with this line of reasoning, any rethink of the traditional roadmap to industrialisation must work towards creating a level playing field for all actors involved in economic development initiatives, be it producers, consumers or other stakeholders.

---

[22] According to the late Dr. Mahbub ul Haq, human security includes seven threats: economic security, food security, health security, environmental security, personal security, community security and political security (UNDP 1994).

5.5.1 China's Efforts on This Front

China has been making efforts toward rethinking its development model for some time. In 2003, for instance, the Chinese government announced that its overall economic and social development strategies would be guided by the concept of scientific development—a concept which refers to the need for China's development to be 'people-oriented', comprehensive, coordinated and sustainable, as opposed to the traditional concept of development which is 'wealth oriented'. In contrast to the 'transformation via economic growth' model in China in the 1990s, the scientific development concept goes beyond the need for efficiency while taking into consideration human development, with emphasis on enabling a harmonious relationship between various stakeholders as well as between stakeholders and the environment.

This initiative was proof of the significant transformation that had occurred, both in principle and in practice, with regard to energy conservation and environmental protection in China. Possibly, in the coming years, these positive changes will come through for China in its development process. A reason for this may be that, in contrast to developed countries, China still has a relatively lighter historical burden of development. In the context of globalisation, China should take advantage of its socialist system and explore further innovative ideas that can help to ensure a consumer lifestyle that somehow remains conscientious of the importance of simplicity and richness in spirituality. Moreover, it should be emphasised that wealth is only a means to an end. China should make efforts to improve efficiency in transforming resources to not only achieve wealth but also the welfare and happiness of its people in spite of the process necessitating some political reform. In so doing, China would enhance social development while using relatively less resources and wealth.

5.6 Conclusion

China's approach towards its energy savings and emission reduction commitments is serious and positive and its efforts have largely been productive. Nevertheless, challenges that may be tedious to overcome do remain. Besides, using energy consumption intensity as an indicator of energy savings, as is the tendency among many in China, is clearly flawed in both theory and practice. The challenges outlined in the chapter point to a need for China to deepen and broaden the concepts of energy intensity and economic development in order to realise its energy saving and emission reduction targets more reasonably and effectively.

Where achieving targets on energy savings and emission reductions is concerned, China should avoid establishing short-term energy consumption targets, given the uncertainty in short-term trends of energy consumption intensity, but give due emphasis to short-term fluctuations while formulating long-term energy saving and emission reduction policies. Given that China's energy use per capita is still below the world average and less than one-third that of developed countries, it should be expected that there will inevitably be a substantial increase in its energy consumption prior to 2020. Nonetheless, it may be possible for China to take on its commitment of effective reduction of carbon emissions after 2010. In order to ensure that circumstances remain conducive to this commitment, it is advisable that China does not pursue a significant and substantial decrease in its energy intensity in the short term—any such decrease in the short term would discourage long-term reductions of energy (and carbon) intensity as well as China's development in the long term.

To ensure sustainable socioeconomic development, it is necessary that China's strategies and efforts toward energy savings and emission reductions be thought through an NTS lens that also effectively alleviates the pressures on its resources and the environment. A more concerted effort is needed to incorporate increasingly innovative options that can improve the efficiency of transforming resources to ensure the wealth and welfare of its communities and societies.

References

BP (British Petroleum) (2010) BP statistical review of world energy, June 2010. BP, London. http://www.bp.com/liveassets/bp_internet/globalbp/globalbp_uk_english/reports_and_publications/statistical_energy_review_2008/STAGING/local_assets/2010_downloads/statistical_review_of_world_energy_full_report_2010.pdf. Accessed 26 Jan 2012

China Statistics Press (2011) China statistical yearbook 2011. China Statistics Press, Beijing

IEA (International Energy Agency) (2007) Oil supply security: emergency response of IEA countries 2007. OECD/IEA, Paris. http://www.iea.org/textbase/nppdf/free/2007/oil_security.pdf. Accessed 29 Feb 2012

IEA (International Energy Agency) (2010) World energy outlook 2010. OECD/IEA, Paris

Pan J, Peng W, Li M, Wu X, Wan L, Zerriffi H, Victor D, Elias B, Zhang C (2006) Rural electrification in China 1950–2004: historical processes and key driving forces, working paper #60. Program on Energy and Sustainable Development, Stanford. http://iis-db.stanford.edu/pubs/21292/WP_60,_Rural_Elec_China.pdf. Accessed 26 Jan 2012

Rawski TG (2009) Urban air quality in China: historical and comparative perspectives. In: Islam N (ed) Resurgent China: issues for the future. Palgrave-Macmillan, Houndmills and New York, pp 353–370

UNDP (United Nations Development Programme) (1994) Human development report 1994: new dimensions of human security. Oxford University Press, New York

UNDP (United Nations Development Programme) (2011) Human development report 2011: sustainability and equity, a better future for all. Palgrave-Macmillan, New York

WWF (World Wildlife Fund for Nature), ECOFYS, OMA (Office for Metropolitan Architecture) (2011) The energy report: 100% renewable energy by 2050. WWF, Gland. http://assets.wwf.org.uk/downloads/2011_02_02_the_energy_report_full.pdf. Accessed 26 Jan 2012

Abbreviations

BP	British Petroleum
CASS	Chinese Academy of Social Sciences
CGE	Computable general equilibrium
CO_2	Carbon dioxide
g	Grams
GDP	Gross domestic product
GHG	Greenhouse gas
gtoe	Billion tons of oil equivalent
IEA	International Energy Agency
IPCC	Intergovernmental Panel on Climate Change
IQTE	Institute of Quantitative and Technical Economics
kW	Kilowatts
kW h	Kilowatts hour
Mt	Metric tonnes
NECC	National Energy Conservation Center, China
NTS	Non-traditional security
NTU	Nanyang Technological University
OECD	Organisation for Economic Co-operation and Development
OMA	Office for Metropolitan Architecture
PBL	PBL Netherlands Environmental Assessment Agency
RSIS	S. Rajaratnam School of International Studies
SME	Small- and medium-sized enterprise
t	Ton
tce	Tons of standard coal
toe	Tons of oil equivalent
UN	United Nations
UNDP	United Nations Development Programme
UNFCCC	United Nations Framework Convention on Climate Change
US	United States
WTO	World Trade Organization
WWF	World Wide Fund for Nature

Author Biographies

Yuxin Zheng (Peoples' Republic of China): research fellow and former Deputy Director at the Institute of Quantitative and Technical Economics (IQTE), and Director at the Center for Environment and Development, Chinese Academy of Social Sciences. His main work areas include industrial productivity, computable general equilibrium (CGE) modelling and policy simulation, and policy studies of the environment and sustainable development. His publications include: (with Rawski, Thomas): *Productivity of China's Industry in Transition* (Beijing: Social Sciences Literature Press, 1993); (with Fan, Mingtai): *China CGE and Policy Studies* (Beijing: Social Sciences Literature Press, 1998); *Economic Analysis of Environmental Effect* (Beijing: Social Sciences Literature Press, 2003).

Sofiah Jamil (Singapore): Adjunct research associate at the S. Rajaratnam School of International Studies (RSIS). Sofiah was previously associate research fellow at the RSIS Centre for Non-Traditional Security (NTS) Studies, where she co-led two programmes—Climate Change and Environmental Security, and Energy Security. She writes regularly for the centre, focusing on policy-relevant perspectives on the environment in the Asian region. She is keenly interested in the role of civil society in human security and environmental issues. Her recent publications include: "Islam & Environmentalism: Greening Our Youth" (*in Igniting Thought Unleashing Youth: Perspectives on Muslim Youth and Activism in Singapore*, edited by M Nawab and F Ali; Singapore: Select Books, 2009). She continues to contribute to the work of the RSIS Centre for NTS Studies while pursuing her PhD (International, Political and Strategic Studies) at The Australian National University.

Energy and Non-Traditional Security (NTS) in Asia: Key Institutions

ASEAN Centre for Energy (ACE)
Indonesia
http://aseanenergy.org/

Consortium of Non-Traditional Security Studies in Asia (NTS-Asia)
Singapore
http://www.rsis-ntsasia.org/index.html

East Asia Institute (EAI)
Republic of Korea
http://www.eai.or.kr/english/index.asp

East Sea Rim Research Center
Republic of Korea
http://erc.khu.ac.kr/eng/default/index.php

East-West Center
USA
http://www.eastwestcenter.org/

Energy Studies Institute (ESI)
Singapore
http://www.esi.nus.edu.sg/

Foundation of Indonesian Institute for Energy Economics (IIEE)
Indonesia
http://www.iiee.or.id/

Gulf Research Center
Switzerland
http://www.grc.ae/

M. Caballero-Anthony et al. (eds.), *Rethinking Energy Security in Asia:* 99
A Non-Traditional View of Human Security, SpringerBriefs in Environment, Security,
Development and Peace 2, DOI: 10.1007/978-3-642-29703-8, © The Author(s) 2012

Independent Power Producers of India (IPPAI)
India
http://www.ippai.org/

Institute for Essential Services Reform (IESR)
Indonesia
http://en.iesr-indonesia.org/

Institute for Public Policy Research (IPPR)
UK
http://www.ippr.org/

Korea Energy Economics Institute (KEEI)
Republic of Korea
http://www.keei.re.kr/main.nsf/index_en.html

Lowy Institute for International Policy
Australia
http://www.lowyinstitute.org/

Nautilus Institute for Security and Sustainability
USA, Australia, Republic of Korea
http://www.nautilus.org/

RSIS Centre for Non-Traditional Security (NTS) Studies
Singapore
http://www.rsis.edu.sg/nts

Samsung Economic Research Institute (SERI)
Republic of Korea
http://www.seriworld.org/

The Energy and Resources Institute (TERI),
India
http://www.teriin.org/index.php

The Institute of Energy Economics, Japan (IEEJ)
Japan
http://eneken.ieej.or.jp/en/

The National Bureau of Asian Research (NBR)
USA
http://nbr.org/default.aspx

World Resources Institute (WRI)
USA
http://www.wri.org/

About the RSIS Centre for Non-Traditional Security (NTS) Studies

The **RSIS Centre for Non-Traditional Security (NTS) Studies** conducts research and produces policy-relevant analyses aimed at furthering awareness and building capacity to address NTS issues and challenges in the Asia-Pacific region and beyond.

To fulfil this mission, the Centre aims to:

- Advance the understanding of NTS issues and challenges in the Asia-Pacific by highlighting gaps in knowledge and policy, and identifying best practices among state and non-state actors in responding to these challenges.
- Provide a platform for scholars and policymakers within and outside Asia to discuss and analyse NTS issues in the region.
- Network with institutions and organisations worldwide to exchange information, insights and experiences in the area of NTS.
- Engage policymakers on the importance of NTS in guiding political responses to NTS emergencies and develop strategies to mitigate the risks to state and human security.
- Contribute to building the institutional capacity of governments, and regional and international organisations to respond to NTS challenges.

Our Research

The key programmes at the **RSIS Centre for NTS Studies** include:

Internal and Cross-Border Conflict

- Dynamics of Internal Conflicts
- Multi-level and Multilateral Approaches to Internal Conflict
- Responsibility to Protect (RtoP) in Asia
- Peacebuilding

M. Caballero-Anthony et al. (eds.), *Rethinking Energy Security in Asia:*
A Non-Traditional View of Human Security, SpringerBriefs in Environment, Security,
Development and Peace 2, DOI: 10.1007/978-3-642-29703-8, © The Author(s) 2012

2. Climate Change, Environmental Security and Natural Disasters

- Mitigation and Adaptation Policy Studies
- The Politics and Diplomacy of Climate Change

3. Energy and Human Security

- Security and Safety of Energy Infrastructure
- Stability of Energy Markets
- Energy Sustainability
- Nuclear Energy and Security

4. Food Security

- Regional Cooperation
- Food Security Indicators
- Food Production and Human Security

5. Health and Human Security

- Health and Human Security
- Global Health Governance
- Pandemic Preparedness and Global Response Networks

Our Output

Policy Relevant Publications

The **RSIS Centre for NTS Studies** produces a range of output such as research reports, books, monographs, policy briefs and conference proceedings.

Training

Based in RSIS, which has an excellent record of post-graduate teaching, an international faculty, and an extensive network of policy institutes worldwide, the Centre is well-placed to develop robust research capabilities, conduct training courses and facilitate advanced education on NTS. These are aimed at, but not limited to, academics, analysts, policymakers and non-governmental organisations (NGOs).

Networking and Outreach

The Centre serves as a networking hub for researchers, policy analysts, policymakers, NGOs and media from across Asia and farther afield interested in NTS issues and challenges.

The **RSIS Centre for NTS Studies** is also the Secretariat of the Consortium of Non-Traditional Security Studies in Asia (NTS-Asia), which brings together 20 research institutes and think tanks from across Asia, and strives to develop the process of networking, consolidate existing research on NTS-related issues, and mainstream NTS studies in Asia.

More information on our Centre is available at www.rsis.edu.sg/nts

Research in the RSIS Centre for NTS Studies received a boost when the Centre was selected as one of three core institutions to lead the MacArthur Asia Security Initiative in 2009.

The Asia Security Initiative was launched by the John D. and Catherine T. MacArthur Foundation in January 2009, through which approximately US$68 million in grants will be made to policy research institutions over seven years to help raise the effectiveness of international cooperation in preventing conflict and promoting peace and security in Asia.